THE PHILOSOPHER
OF PALO ALTO

The Philosopher of Palo Alto

MARK WEISER, XEROX PARC, AND THE ORIGINAL INTERNET OF THINGS

JOHN TINNELL

The University of Chicago Press
Chicago and London

The University of Chicago Press, Chicago 60637
The University of Chicago Press, Ltd., London

Published 2023
Printed in the United States of America

32 31 30 29 28 27 26 25 24 23   1 2 3 4 5

ISBN-13: 978-0-226-75720-9 (cloth)
ISBN-13: 978-0-226-75734-6 (e-book)
DOI: https://doi.org/10.7208/chicago/9780226757346.001.0001

Library of Congress Cataloging-in-Publication Data

Names: Tinnell, John, author.
Title: The philosopher of Palo Alto : Mark Weiser, Xerox PARC, and the original Internet of things / John Tinnell.
Description: Chicago : The University of Chicago Press, 2023. | Includes bibliographical references and index.
Identifiers: LCCN 2022045649 | ISBN 9780226757209 (cloth) | ISBN 9780226757346 (ebook)
Subjects: LCSH: Weiser, Mark. | Xerox PARC (Firm) | Internet—Social aspects. | Digital communications—United States—Biography. | Computer software industry—United States—Biography.
Classification: LCC HD9696.63.U62 W4585 2023 | DDC 338.7/610053092 [B]–dc23/eng/20220923
LC record available at https://lccn.loc.gov/2022045649

⊗ This paper meets the requirements of ANSI/NISO Z39.48-1992 (Permanence of Paper).

for my parents

CONTENTS

PROLOGUE

NICHOLAS NEGROPONTE, head of the storied Media Lab at the Massachusetts Institute of Technology and patron saint of *Wired* magazine, is lecturing about the future of technology to an auditorium packed with computing's brightest minds. An actor dressed in a butler costume stands near Negroponte on stage, pantomiming his remarks. The gimmick is meant to illustrate Negroponte's belief that the next wave of computers should be like butlers—artificially intelligent aides ready to take our orders and serve up whatever information we wish. Known as "interface agents" in technical circles, these digital assistants will eventually recognize human speech and respond in kind, Negroponte tells the crowd. They have gathered here together for the world's first symposium on interface agents. It is October 20, 1992. In the coming years, Negroponte will soon write an influential bestseller and later spearhead an initiative aiming to give every kid in the world a laptop, all while running one of the most lucrative research laboratories in the history of academia.

Not far from Negroponte and the butler, waiting for his turn to speak, sits a balding man in his forties wearing red suspenders over a baggy dress shirt, a look that is no more fashionable in 1992 than it would be today. His name is Mark Weiser. Despite his rumpled appearance, he exerts a considerable influence, too, and his star is rising. Technology reporters have begun flocking to visit Weiser's office in the Computer Science Lab at Xerox's Palo Alto Research Center (PARC), eager for a sneak peek at the twenty-first century. Upon seeing the devices on display there and hearing Weiser talk about them, one reporter will conclude that "Mark Weiser might rearrange society as thoroughly as Thomas Edison did when he electrified the cities."[1] Years later, another reporter, charmed by Weiser's philosophical motivations, will deem him "the soul and conscience of Silicon Valley."[2]

Weiser has made this trip to MIT knowing that the day's symposium might not go well for him. For one thing, he disagrees with Negroponte. He also disagrees with Alan Kay, the computing icon who held court on stage before Negroponte. In fact, he largely disagrees with each of the other eight speakers with whom he's been invited to share the podium. Once it's his turn, Weiser intends to challenge the very idea that the symposium is meant to advance. He plans to argue that interface agents—the digital assistants personified by Negroponte's butler—represent a massive step in the wrong direction.

Up until the previous year, many of the symposium's six hundred attendees had likely never heard of Mark Weiser. That changed, almost overnight, when an article he wrote for *Scientific American* appeared in the magazine's September 1991 issue, alongside pieces by Negroponte and Kay. Weiser's article—"The Computer for the 21st Century"—went viral, in the way things went viral before social media. Bill Gates read it and immediately dashed off a memo to his Microsoft executives insisting that "everyone should read" it.[3] Postcards from computer scientists around the world poured into Weiser's office, each asking him to have Xerox mail them a free copy. Soon the *New York Times* and *Washington Post* lent more hype to the big idea his article had put forth: the next great leap in the evolution of computers, Weiser suggested, will be their disappearance. Desktops and laptops will gradually be overshadowed by billions of smaller devices, and these won't look like computers at all. The power of computing will be built into all kinds of familiar objects: clocks, coffeepots, pens, doors, windows, car windshields, and many more. Everyday things around the home, the workplace, and urban spaces will become seamlessly infused with connectivity.

The broad strokes of this vision—Weiser calls it "ubiquitous computing"—have landed him speaking roles at many venues like the

MIT auditorium he is sitting in. And now that he's part of these conversations, it's the details that concern him most. He tells his colleagues that computers must be rethought and rebuilt—piece by piece—if they are to truly "fit the human environment instead of forcing humans to enter theirs."[4] Personal computers like desktops and laptops exist, Weiser says, "largely in a world of [their] own."[5] It is a world he knows well. Ever since he learned to program on punch cards at age fifteen, Weiser has been captivated by the machines he spent his life mastering. He knows exactly what Steve Jobs means when he says the computer is a bicycle for our minds that "can amplify a certain part of our inherent intelligence."[6] Still, Weiser harbors mixed feelings about computers. For all the ways PCs augment our thinking, he worries that they hold our bodies captive, that they demand too much of our attention and ultimately weaken our connection with the wider world off screen. He believes that adding interface agents to the mix—those chatty, watchful digital assistants that Negroponte and the others are championing—could actually make matters worse.

Back at Xerox PARC, Weiser and his colleagues have created a slew of handheld gadgets that connect wirelessly to one another throughout their building. They have crafted mobile software that automatically displays content related to their location and shares their whereabouts with friends in real time. Other applications sync with the building's energy systems to precisely adjust each room's lighting and temperature as the staff comes and goes. But even these inventions already feel to him like stepping-stones on the path to something else—they are "phase I," he says.[7] He's searching, from one experiment to the next, for a more graceful means of weaving digital information into our surroundings, of mobilizing computational resources to "increase our ability for informed action" as we move about the world.[8] The metaphors he uses to describe the human-computer interactions he craves are hard to sell and hard to build. "Our computers should be like our childhood," he thinks: "an invisible foundation that is quickly forgotten but always with us, and effortlessly used throughout our lives."[9] He knows that his loftiest goals are moon shots, but he maintains that continued technical advances will render them viable someday. Developing a philosophy to guide the technology's design has become his chief obsession. He estimates that his vision—"ubiquitous computing"—will be "a twenty-year quest."[10]

Inside the MIT auditorium, it is now his turn to address the crowd. The butler introduces Weiser and calls him to the stage. Thumbing his suspenders, Weiser takes the podium.

This is the fourth year of his twenty-year quest. The future he imagines for technology—for himself—will not unfold as planned.

INTRODUCTION

Googleville

> Do we really think that everything in the world would be better if it were smarter? Smart Cappuccino? Smart Park? The "Smart House" of 2005 will have computers in every room. But what will they do?
>
> Mark Weiser, "Open House," 1996

MORE THAN ONE HUNDRED BILLION THINGS are expected to be connected to the internet by 2030.[1] Among them are thermostats that learn and remember, cars that navigate and intervene, doorbells that observe and alert, and mattresses that calculate and self-adjust. Ever since cell phones became smartphones, technologists and consumers in high-tech societies have come to embrace all manner of smart objects with astonishing speed. Connectivity has spread rapidly from personal computers into parking lots, toilets, eyeglasses, and kitchens. Roughly ten thousand websites were online in 1994; soon, appliance manufacturers expect to sell two million Wi-Fi-enabled refrigerators per year.[2]

Over the past fifty years, this once-eccentric longing to animate the inanimate has become a global enterprise. Computer scientists have progressed quickly from the first bona fide "connected device" in 1970 to twenty-first-century designs for wholly connected "smart cities." Technologists often trace the Internet of Things' humblest beginnings to a Carnegie Mellon University soda machine, which had been rewired by professors who wished to monitor its exact contents from their offices. Carnegie Mellon's Computer Science Department reportedly drank "120 bottles of Coca-Cola products each day" in 1970; too often, the faculty found themselves ascending the stairs to their building's third floor only to find the soda machine empty.[3] It is hardly a stretch to see a reflection of that empty soda machine in the problems that more recent Internet of Things devices aim to solve. Many so-called smart products hitting the market remain devoted to tackling life's trivial irritations. In his 2015 *Atlantic* essay "The Internet of Things You Don't Really Need," Ian Bogost lampooned his pet case in point: a mobile app-sensor system called "Smart GasWatch" that leveraged Bluetooth technology to allow its customers to check the level of their grill's propane tank *using their smartphones*, rather than a mechanical gauge or several other cheaper means. "Today," Bogost wryly observed, "the relevance of any consumer product requires the addition of superfluous computing."[4] Household gadgets have traded largely on our eagerness to be always within a glance of digital metrics that help us feel more in command, even when eyeing this data makes little difference.

But the race to render everything smart—sensor laden, data rich, instantly adjustable—has moved far beyond the stuff of backyards and bedrooms. The most ambitious North American plan to date surfaced in October 2017, in Toronto, where city officials launched a partnership with Google's Sidewalk Labs to create a city-within-the-city "built from the Internet up," on a languishing industrial stretch of waterfront called Quayside.[5] Sidewalk Labs' proposal laid out a thoroughgoing merger between urban infrastructure and new tech. Enhanced sidewalks would instantly heat away snow while also gathering information about everybody who traversed them. An AI-driven electrical grid and autonomous transit system would give off 89 percent fewer greenhouse gas emissions than Toronto's existing neighborhoods. Nothing would go unmeasured or unmonitored. Inefficiency, waste, and discomfort would be minimized with religious zeal. Google cofounder Larry Page and Canadian Prime Minister Justin Trudeau both insisted that an omniscient city could radically boost its residents' quality of life. Trudeau was especially bullish, stating at the project's onset: "I have no doubt Quayside will become a model for cities around the world."[6] Improvements across all municipal services that might otherwise take months or years to discuss and implement would roll out swiftly and continuously thanks to the new wellsprings of citizen data that Sidewalk Labs pledged to unearth and analyze.

Unsurprisingly, the project's announcement raised concerns from a slew of critics worried about the fate of human agency in such an environment. As urbanists, academics, and journalists combed through the proposal, they developed a bleak perspective on the big picture. At the same time, Google began to make greater demands. The twelve-acre Quayside parcel that it initially agreed on suddenly appeared too small. Sidewalk Labs, with Google's backing, pushed to expand the footprint to 165 acres, giving them room to build a new Canadian Google headquarters and reason enough to allocate Can$1.3 billion. Much of Toronto interpreted the move as an act of aggression, and the critics aired their objections to receptive ears. "In Sidewalk Labs' scheme, residents provide (unpaid) feedback about the products they use—but without gaining any political agency in return," noted the architecture professor T. F. Tierney. "There is no consideration of context, no opportunity for expression or deliberation or debate. Data decides."[7] If closed-door calculations were to take the place of public discussion, the resident inhabiting Toronto's Googleville would no longer be a citizen with a voice. Instead of using advocacy and dialogue, she would represent herself by allowing the bulk of her existence to be tracked by tiny sensors and computer-vision-powered cameras. It would be her civic duty to be a good human dataset, allowing each of her activities to be tabulated and processed by the appointed algorithms entrusted with making the place better, or at least more efficient.

Some of the plan's more disconcerting aspects were not entirely futuristic. For several years already, advances in "insurtech" had brokered a data-driven relationship with insurance policyholders that predated Googleville's envisioned social contract. In 2014, the *New York Times* ran a story about "subprime" borrowers who consented to have a "starter interrupt device" installed on their vehicles in order to retain their auto insurance.[8] This gadget enabled insurers to keep tabs on the borrower's vehicle at all times, even granting them power to remotely shut down the vehicle's engine in response to observed policy breaches. The story caught the attention of Jathan Sadowski, a technology scholar who proceeded to examine all the subtle ways corporations were beginning to experiment with smart devices. Akin to the auto-starter interrupter, there was a smart toothbrush sold by Beam Technologies that captured data about its users' every brushing (duration, performance rating, time of day) and transmitted it to Beam's dental insurance staff. After taking inventory of many such practices cropping up in a wide range of sectors, including advertising and policing, Sadowski had heaps of evidence lending dramatic weight to his general conclusion: "With detailed data monitoring comes the power of behavioral modification."[9]

Behavioral modification was, however, the very quality that Toronto's most ardent Sidewalk Labs supporters celebrated in their pleas to move forward in the fall of 2019. Kwame McKenzie, the CEO of Toronto's Wellesley Institute, lauded Sidewalk's vision for Quayside as "a place to prototype ideas" that may well "produce scalable strategies that Ontario could use to combat climate change."[10] For him, the detailed blueprint of a "low-carbon, resilient neighborhood" with locally sourced, Google-funded affordable housing was not something to shrug off, even as he acknowledged the lack of data safeguards: "We could focus on issues such as privacy while seemingly forgetting that our current economic, social and health problems are a matter of life and death for some."[11] The privacy expert Ann Cavoukian was more intent on establishing safeguards, as she had witnessed firsthand how little Sidewalk's assurances were worth. She agreed to serve as a consultant for the Quayside development at the company's request, but only on her condition that "any data collected must be de-identified and anonymized immediately at the source."[12] Sidewalk agreed, until they decided to create the Urban Data Trust, a group consisting of multiple tech companies and government entities, which they billed as an institution to democratize access to Quayside data. That sounded okay to Cavoukian—and then the fine print was revealed. "Sidewalk also decided that it would not, or could not, compel other members of the trust to de-identify their data," Cavoukian learned. "They said they'd 'encourage' them."[13] She resigned from her position and doubled down publicly on her call for data privacy.

Of all the critics who scrutinized Google-Sidewalk's plan for Toronto, none was more damning than Shoshana Zuboff. She deemed the city a "new frontier" of *surveillance capitalism*, a term she had recently popularized to illuminate the financial ploys underpinning Big Tech's spectacular growth. Google—the veritable pioneer of surveillance capitalism, according to Zuboff—began the practice in the early 2000s with targeted advertising tools called "AdWords" and "AdSense," which mined users' keywords and clicks in attempt to "read users' minds for the purposes of matching ads to their interests."[14] Google then greeted the rise of smartphones with its Android mobile OS, its Maps navigation app, and Pokémon GO—all of which supplied deeper insights into users' psychology and improved the company's ability to observe and predict behavior, as well as nudge it on location to benefit their high-paying advertiser-clients. Sidewalk CEO Dan Doctoroff announced plainly that they would extend this approach to all "Google City" projects: "We fund it all . . . through a very novel advertising model. . . . We can actually then target ads to people in proximity, and then obviously over time track them through things like beacons and location services as well as their browsing activity."[15] Under this model, whatever technological possibilities the Internet of Things might weave into the urban fabric would inevitably tilt toward corporate, consumerist ends. Promised improvements to Torontonians' quality of life could turn out to be secondary goals at best or, at worst, "hooks" that served primarily to "lure users into . . . extractive operations."[16] After citing Google's consistent track record, Zuboff warned of a looming domino effect: "Should Toronto fall to this anti-democratic juggernaut, surveillance capitalism will be emboldened to keep on taking. . . . More cities, then regions, then countries will be reborn as private data flows that yearn toward totality for the sake of profits, until 'privacy,' 'self-determination,' and 'democracy' read like ancient words on faded parchment."[17]

On May 6, 2020, Sidewalk Labs pulled the plug. While two years of sustained public backlash no doubt weighed into the decision, Doctoroff said it was the "economic uncertainty" brought about by COVID-19 that convinced the firm to abort their plans.[18] Ardent detractors celebrated their victory over Google on social media; local fans of the project typed out mournful comments about what could've been. But news of the project's termination passed largely unnoticed as the world struggled with the pandemic. A year later, Toronto mayor John Tory revealed his take on the botched deal, claiming that the uproar about data privacy wasn't the pivotal issue—real estate was. Sidewalk expected to get a "bargain-basement price" on the land because of the pandemic, according to Tory, but the city did not give in.[19] At any rate, the sudden turn of events in Toronto reads more like an ellipsis than a resolution.

Attempts to create some kind of Googleville—an ambition long held by Larry Page and others of his ilk—will almost certainly continue. Bill Gates has almost twenty-five thousand acres near Phoenix reportedly earmarked for smart-city construction. The e-commerce billionaire Marc Lore announced his intentions in 2021 to get into the game. Other companies are progressing on their own initiatives around the world: Panasonic's Sustainable Smart Town in Fujisawa, Japan; Masdar's Masdar City in Abu Dhabi; and Cisco's development of Songdo, South Korea, not to mention the onslaught of IoT-enabled amenities and services that have made a place for themselves just about anywhere with decent Wi-Fi. The battle over Toronto has subsided for now, but more tech moguls with similar aspirations are longingly browsing world maps.

The Internet of Things is still a spectrum of prospects to be reckoned with. For all the billions of new gadgets coming online, our current debates about what they should and shouldn't do fail to present a set of common ideals we might work toward. We stand guarded against the vivid Orwellian nightmare, while talking past one another's desires in the absence of a shared dream. Proponents of smart infrastructure are right to enthuse about climate solutions, improved health outcomes, and research breakthroughs that could result from dense networks of connected objects. Even some critics—Zuboff included—hold out hope for some alternative future in which the Internet of Things (and the wealth of data it generates) "becomes a critical resource for people and society," rather than a top-down tool for surveillance capitalists, attention merchants, and remote technocrats.[20]

So how might we bring about this better future? In addition to regulating our way out of the present, we must rally around ideas that promise more for the public good. We can find such concepts not only in the recent manifestos of tech-savvy architects and urbanists, but also in the dustbin of computer history. Prior to the Googleville model, another approach had been incubated elsewhere in Silicon Valley.

Like a forest growing up from acorns, the basis for smart cities and homes began inside just a few smart rooms occupying a handful of buildings. The practices that came to be favored and embellished by Google had initially emerged, as Zuboff has argued, from the MIT Media Lab, "where some of surveillance capitalism's most valuable capabilities and applications, from data mining to wearable technologies, were invented."[21] This lineage actually traces further back than her seemingly exhaustive analysis powerfully uncovers. Its history feeds into a lesser-known origin story set in the 1990s. Much of the technology forged at MIT during the early years of the new millennium had already been developed, quietly and quite differently, a decade before in Palo Alto, California.

When I began my research for this book in the prodigious paper enclave of Stanford University's Special Collections room, hunching daily over worn documents and surrounded by other writers doing the same, I never imagined a sunny afternoon in March when I'd be the lone patron.

I had come to Stanford's archives to see what the Internet of Things looked like during the 1990s, back when it was just a glint in the eyes of a few researchers. Silicon Valley was busy realizing other visions then. Desktops and laptops were finally achieving mass adoption; newfangled mobile devices like the Newton, Apple's first handheld computer, had failed to catch on. The World Wide Web sprang up from its academic roots and blossomed into the dot-com boom. As tech CEOs and venture capitalists leaped into cyberspace, the effort to computerize everyday things in physical settings launched under relative obscurity. I wanted to examine these early unsung projects in fine detail to better understand how they compared to current smart devices, IoT systems, and smart-city plans. In these formative years before surveillance capitalism took hold, what was the inciting mission behind the original Internet of Things? Why did its most outspoken champion dedicate himself so doggedly to creating internet-enabled tablets, pens, coffeepots, office windows, walls, and ID badges when all the smart money was betting on personal computing and virtual reality? What kind of a future did he and his collaborators hope to lay the foundation for in their lab? Perhaps their pioneering work would hold some conceptual tenets for building a better Internet of Things today. Breakthrough moments in computer history sometimes constitute "a richer past for our cruder present," as the digital luminary Alan Kay likes to remind audiences of younger technologists.[22]

My days in the archives soon began to yield more than I expected. In addition to finding answers and insights, I stumbled upon a story. At the heart of this intellectual quest to reimagine the purpose and place of digital connectivity was a subplot about one innovator's lifelong struggle to *feel* connected to the world.

On Wednesday, March 11, 2020, I sat alone in the Special Collections room, rushing through boxes full of lab notebooks, PowerPoint slides, folded letters and printed emails, unpublished stories and poems, quarterly reports, staff evaluations, business plans, and project proposals with odes to failure scribbled in their margins. Adrenaline eviscerated any thought that wasn't pertinent to the task. For a dire morning, I became an archival-research machine. My eyes darted from page to page; as my left hand turned the pages, my right hand took photos as fast as it could. When I had woken up a few hours earlier, I still believed I was only halfway through a semester-long stay. Everything had become uncertain in the hours since, and I figured it might be my last day there for a while. Multiple cases of COVID-19 had just been confirmed on campus. Buildings hadn't yet been shut down, but everyone wondered when they would be. The World Health Organization deemed the situation a global pandemic later that day.

The motive behind my urgency was simple, personal, and a bit irrational. I was unraveling the narrative threads of a thirty-year-old story, enthralled by a dead man's collected papers that weren't going anywhere. They'd be waiting for me at Stanford afterward, and their relevance to the technological horizons of twenty-first-century life would only be amplified by the pandemic, when remote work and video calls would become the norm. The old documents offered a rare glimpse into an untold chapter in this history of Silicon Valley, one set in the laboratories of its most revered R&D site, Xerox's Palo Alto Research Center. Paragraphs and pictures filled the gaps in a film that was forming in my head, nearing its climax, and I wasn't going to stop until the archivists told me to leave.

In my mind's eye, I saw how the seeds for the Internet of Things had been sown before the onset of the World Wide Web, and how a group of Xerox PARC researchers had turned their headquarters into a tiny little smart town long before the first iPhones and iPads were conceived. Equally intriguing as PARC's technologies was its cast of lead characters: Mark Weiser, PARC's chief technologist who loved existentialist philosophy and scorned personal computers, even though he used them constantly; Lucy Suchman, an anthropologist who pioneered the ethnographic study of human-computer interaction; Rich Gold, an ex-musician/artist-turned-toymaker who had a reputation for asking the strangest questions anyone in Silicon Valley ever heard; and John Seely Brown, the director of PARC, who generally steered talk at meetings away from budgets and toward theoretical debates about the nature of knowledge—and who would throw his Birkenstock sandal at you if he thought your ideas were either really good or especially bad.[23] What was truly captivating about these PARC researchers, in addition to having incubated several influential devices, was their commitment to being philosophers of the technologies they invented. The conversations they were having—be it about connectivity and control, the merits and limits of artificial intelligence, or the ideal role of mobile interfaces in daily life—were prescient for a reason: throughout the 1990s, Weiser and his computer science colleagues turned PARC into a showroom for a future whose threshold we are just beginning to cross.

Xerox PARC has been rightfully hailed as a birthplace of personal computing for its trailblazing work in the 1970s. Its scientists molded almost every building block that rendered desktop machines user friendly to the masses. PARC in the '70s served, among other things, as the indispensable bridge that transported grand notions from academia and advanced them toward mainstream adoption, making it easier for Apple and other upstarts to focus largely on marketing this new kind of computer to layperson consumers as their engineers added the finishing touches. While so much has been made about that golden age spanning PARC's opening years (as well as Xerox's ensuing failure to capitalize on its researchers' inventions), decades of subsequent innovation at Xerox PARC remain overshadowed by the lore surrounding that initial rise and alleged fall. The history of Xerox PARC, in the minds of tech enthusiasts, journalists, and even some historians, simply ends in the mid-'80s, when PARC notoriously "fumbled the future" of personal computing.[24]

But it became wildly evident to me (and to the five or six other scholars who had poked around in the archive) that Xerox PARC had a renaissance in the '90s. A second golden age flourished then that may prove every bit as impactful as the first, though it is difficult to measure fully, since the ripples of PARC's work in the '90s are still being felt in the new waves of technology we are now confronting. Today's innovators are still learning from this legacy, and Weiser's expansive view of the internet continues to catch on in technical circles. Apple, Amazon, Microsoft, and Google are vying to make all things smart, from watches and glasses and cars to grocery stores and cities. Google cofounder Sergey Brin seemed to paraphrase one line after another from Weiser's 1991 article ("The Computer for the 21st Century") in his 2013 TED Talk demoing Google Glass, as did the company's former CEO, Eric Schmidt, in 2015 when he told the World Economic Forum in Davos that "the Internet will disappear."[25] After a dramatic pause, Schmidt explained: "There will be so many IP addresses *because of iPv6*—so many devices, sensors, things that you're wearing, things that you're interacting with—that you won't even sense it. . . . [The internet] will be part of your presence all the time."[26] Schmidt and Weiser had sat next to each other on a panel at the 1998 World Economic Forum, where Weiser had captivated the audience with his talk of technologies that would blend naturally into everyday life; IPv6—the network protocol that has enabled the Internet of Things to grow—was partially developed in the lab Weiser managed at PARC. The MIT computer scientist Joseph Paradiso traces much of the internet's recent evolution back to Weiser's foresight, going so far as to suggest that Weiser's pioneering essays and inventions should be ranked among the most prescient and seminal contributions in computer history.[27]

And yet, outside of tech's thought leadership circuit, the original IoT vision that Weiser championed at PARC has never quite surfaced in the mainstream. Some of his finer points have faded over the last two decades. Even professionals in the know, who allude to his notion of "ubiquitous computing" with an almost spiritual reverence, sometimes evoke the term in ways that unwittingly contradict the details of Weiser's own writings, which are rarely taught anymore. Appalled at the ironies, the renowned tech designer Amber Case has insisted that Weiser, Brown, and Gold "were so far ahead [of their time] that their work is in danger of being forgotten precisely at a time we most need it."[28] Such was the call I was trying to answer in Stanford's desolate library.

Something changed on March 11, 2020, if only for a few days. Silicon Valley's frenzied sprint to upgrade and outperform fell silent as COVID-19 came into its local communities. Amid this hush I could hear, for the first time, murmurs pining in the voices of those old documents. Here the narrative bore deeper than the one I had plotted from a distance. The anticipated contours of a success story—of a triumphant arch of influence spanning from PARC's trailblazing prototypes to our budding galaxy of connected objects—started to sag and dim as I read on. Just as reading Shoshana Zuboff's work had exposed an unseen layer of smart-city plans, reading through Weiser's personal papers revealed a gulf separating his initial hopes for our digital future from what's currently happening. His pursuit of a seamlessly connected life was riddled with misunderstandings and missed connections; the devices he desired and created came to inspire other devices he dreaded. When I spoke with the people who knew Weiser best, they made clear what I was just starting to suspect. They said that his sudden and much too early passing in 1999 had a silver lining in retrospect: at least he didn't have to see what's becoming of the technologies he championed, for the sight of it all would have broken his heart.

The story of Weiser's time at PARC debunks any notion that technocratic manipulation—total surveillance and zero privacy, runaway automation, and diminished agency—is the inherent cost of living with the Internet of Things. Big Tech's exploitative data practices and covert revenue streams were manufactured out of flagrant disregard for the philosophy that inspired the machinery. Miles apart from Mark Zuckerberg's motto "Move fast and break things," one of Mark Weiser's pet mantras was "Start from the arts and humanities."[29] He and his collaborators supplemented their tinkering sessions with heady discussions drawing on ideas in anthropology, psychology, architecture, phenomenology, science fiction, sculpture, feminism, and the history of writing. Their shared mission was to cultivate a long view that linked R&D to the ethical questions and existential quandaries that each new prototype introduced. For instance, can we connect everything to the internet without constantly taxing our cognitive resources? Against tech's fervor for bells and whistles, PARC scientists aspired to create what they dubbed "calm technology." At the same time, they aspired to deploy sensors and software without rendering the individual less valuable. So-called smart objects, they insisted, should first and foremost make each of us smarter; tech that made people superfluous was the dumbest of all. The scientists would use whatever they built in an effort to discover the unintended consequences and share their concerns with the public. "We must have the will," Weiser would write, after Suchman and Gold opened his eyes to dystopian prospects, "[to] firmly establish a right to privacy of all personal information on any computer, no matter who owns the machine."[30] While the computer industry rode the dot-com boom, PARC researchers were challenging one another to reconcile the dilemmas that innovators so often ignore. They warned readers that a troubling "attention competition" was growing up around the late-'90s web, and that this battle for eyeballs was primed to get ugly as the internet's reach expanded. PARC was arguably the only corporate lab in Silicon Valley where the big picture mattered more than the bottom line. Learning from their struggles to get it right remains our best starting point for building a better Internet of Things.

I should lay bare a sentiment I've absorbed from this archival material, one that has guided my efforts to give this material its shape. Normally it's safe to assume that armies and businesses dictate the formative stages of technical invention; GPS navigation, for example, was devised to secure advantages for US soldiers on desert battlefields in the Persian Gulf, then later outfitted for civilian use by companies seeking to amass users to please their shareholders. The one-two punch of war and profit molds so many gadgets. But looking at invention squarely through an institutional lens can numb us to the undercurrent emotions—stubborn pain, ancient anxieties, impossible wishes—that can grip technologists from adolescence onward and compel their life's work more surely than any militaristic or market force. The history of GPS, like that of maps and compasses before it, goes astray whenever it loses sight of the terror surging in a lone traveler the instant he realizes he is lost. Sometimes the desire to invent comes from a drive to answer fears and longings that never abate. Before combat and after commerce, there is the recurrent white noise of a mind conflicted, the lingering images of love, and of a loved one's passing. A killer app can be a rewarding afterthought, a handy cover-up, a means to justify publicly this most private pursuit. "The machine," wrote the historian Lewis

Mumford, "is just as much a creature of thought as the poem."[31] Any genuine vision of technology begins as a dream for humanity, cast out from the shared miseries of existence that gurgle even in our happiest of happy times. At least, that is how it was around parts of Xerox PARC during the '90s, and too few other places since.

1

Messy Systems

IT HAD ALWAYS BEEN a difficult place to describe. Neither a tech company nor a government lab, partly a center for pure discovery and partly a crystal ball for its parent corporation, Xerox PARC hid in the woodsy foothills of Palo Alto like the technical wing of some impossibly bright university whose researchers had been granted an indefinite break from teaching.

It was a daytime dormitory that housed PhD holders during their waking, working hours. World-class scientists dressed in whatever they pleased would roam the gray-carpeted hallways, popping their heads into one another's offices, airing sudden ideas. They lounged together on beanbag chairs, talking over their ambitions late into the night.

If you walked the hallways long enough, you'd probably overhear more-troublesome conversations that smacked of a frat house. A few AI specialists would regularly huddle around a break-room coffeepot, stroking their assorted facial hair as they ruminated over how to "build a Laura"—that is, a sentient machine modeled after the cheery front desk receptionist, Laura, who greeted them every morning without knowing about their fantasy to automate her. And while geeky male chauvinism was frequently on display, PARC was also home to a notable pantheon of women researchers, owing partly to its multidisciplinary bend.[1] It hired an array of experts not only within the hard sciences—like the blending of physics and chemistry that passed for radical at Bell Labs—but across a loose spectrum of fields that just might prove vital in their crusade to invent the future. Interface designers lunched next to anthropologists; computer programmers smoked on the patio with acclaimed linguists and bona fide philosophers. Even artists could be spotted with some regularity.

Along this spectrum, the bulk of PARC's two-hundred-person research staff was united by its members' impeccable pedigrees. Many had devoted their twenties to knocking out a dissertation at MIT, Stanford, or Carnegie Mellon only to spend their thirties gathering further accolades at some other prestigious lab, all in the hopes of someday claiming an office in this temple of a building overlooking Silicon Valley high atop Coyote Hill Road. Xerox PARC was the place where you could make your name alongside the biggest names.

John Seely Brown saw no need to reign in the intellectual expanse, now that he was mostly in charge. If anything, there was even more he'd like added to the mix. It was the summer of 1987, and most everyone at PARC had by now taken to calling him JSB. He had been promoted the year before from manager of the Intelligent Systems Lab to vice president for advanced research; in two years, just prior to turning fifty, Brown would be named PARC's new director. This was not entirely unexpected, nor was his unusual approach. His extraordinary aptitude for math drew recognition early and often during his boyhood in Hamilton, New York—a remote village dotted with crops and livestock. Before making his way to Brown University, he had passed adolescent summer days working the land and tending cows and sheep in North Adams, Massachusetts. "The farm work opened me up to thinking about the mess, to dealing with messy systems," Brown recalled.[2] In messy systems, as in wicked problems, constantly moving parts and tangled feedback loops gave rise to levels of escalating complexity that often muddled the attempt to pinpoint cause and effect. The pursuit of innovation at PARC often reminded him of the farm in that way. However stately its intricately patterned three-story facility appeared to those outside, PARC was on the inside an exquisite mess held loosely together by a lofty mission.

Back in the summer of 1970, Xerox executives in Rochester, New York, had issued PARC's founding director, George Pake, a mandate to create "The Office of the Future," replete with whatever technologies one might find there.[3] By funding PARC's long-shot efforts, the photocopier giant hoped to give itself a head start on building new devices that might eventually complement or compete with their flagship machine. Before the decade's end, the first generation of PARC researchers delivered a series of remarkable inventions that lay the foundation for personal computing: the first laser printer; Ethernet technology; and a revolutionary programming language called Smalltalk (which powered an even more revolutionary desktop called the Xerox Alto), just to name a few. The future Google CEO Eric Schmidt began his career as a young researcher at PARC around the time of Steve Jobs's visit there in December 1979, during which Jobs got his first glimpse of the graphical user interface that he and his engineers would quickly build into Apple's subsequent personal computers, the Lisa and the Macintosh. "I don't think it's possible to overstate the impact Xerox PARC has had on computing," Schmidt would later say. "Microsoft and Apple get all the credit, but it should go to Xerox PARC."[4] By the mid-'80s, many of the best minds from PARC's original cohort were leaving to go work for other companies that had beaten Xerox to the computing market. Just as the promise of personal computers had become clear, PARC was losing its initial sense of direction.

Nonetheless, PARC was still the dream job for young computer scientists around the world who were drawn to its eccentric lab culture. And Brown knew how to sell it to top recruits. Almost as significant as PARC's innovations during the 1970s, at least in his eyes, were the homespun approaches to R&D that cropped up alongside them. The working principles and processes that emerged then still resonated. They could be mobilized to forge a whole new vision once more. In keeping with PARC tradition, Brown believed it was not the job of the leadership to say what that vision might be—though of course he had his predilections. The center's directors and managers had always resisted making top-down mandates in favor of letting ideas bubble up from the research staff. What captivated Brown was "the white space between fields," where something utterly new might emerge.[5] His only mandate was to hire the best possible people regardless of academic discipline, push them to converse, and then find ways to keep those conversations going.

During his nine years at PARC, Brown had learned much from his administrative predecessors that he wished to preserve and build upon in his new role. He considered the absolute commitment to research freedom a key asset for the center moving forward. PARC scientists enjoyed an uncommon latitude to pursue their own interests, without much concern for government funding or short-term profits. Invariably, it seemed, this was what everyone loved about working there. It was a luxury paid for by Xerox's stronghold on the global photocopier market. Bert Sutherland, the first lab manager Brown worked for after joining PARC in 1978, instructed his staff not to bother with applying for grants or rushing to ship products. "We've got a big mission here," Sutherland told Brown and the others. "We're trying to figure out how to build technology that helps people become more productive. And, no, I don't know what that means, John—you go figure it out."[6] The steady adoption of hardware and software in offices and homes, which gained considerable momentum through the 1980s, gave the scientists the impression that their labs stood at the frontier of a new era. The big mission kept on feeling bigger. "There was just a sense that PARC could be the most fantastic opportunity to build a better world," or even "a new way of being," said Brown.[7] Leaving the center's agenda open to each recruit's interpretation ensured a steady state of mess, and you needed to have a mess in order to have a messy system.

Wrangling these cerebral passions into fruitful collaborations was the objective behind PARC's other core tenets and rituals. Bob Taylor, the legendary manager who was one of Pake's earliest hires, had developed ways to summon researchers from their respective silos into weekly forums that forced them to explain themselves to one another. Taylor's Computer Science Lab—CSL—gathered every Tuesday at 11 a.m., when each week a different scientist took the stage for an hour and held up their latest project to a firing squad of questioning from all angles. One after another, each CSL researcher took a turn in the weekly spotlight and labored to persuade the others on the merits of his or her pet project, in hopes of winning additional resources and willing collaborators. Their colleagues in the audience were both judge and competitor with a vested interest in lobbying for their own initiatives. For every lab member who showed up eager to be wowed, there were others who came ready to attack at the slightest opportunity.

Strange, horrifying things sometimes ensued. Computer scientists reportedly sprang from their seats and marched right up to the day's speaker so as to make a spectacle out of interrupting the presentation with a rebuke that simply could not wait. The beanbag chairs that Taylor famously brought into a PARC conference room—which visiting journalists often cited as symbols of CSL's playful, hippie-leaning spirit—were really just a tool for managing conflicts. "It was impossible to leap to your feet and denounce someone from a bean bag," PARC icon Alan Kay would explain.[8] The Darwinian brand of in-your-face peer review spread from Taylor's meetings to the PARC-wide "Thursday Forums." "Some people didn't want to present within PARC because it could be so hard," recalled John Maxwell, who began his long career there in 1978. "The joke was . . . you should first present [your work] as a keynote talk at an international conference—that would prepare you for the PARC audience."[9]

Brutal as it was, this atmosphere of endless debate pressured PARC researchers to talk to one another across terminological divides. It also fueled all-night tinkering, since the best way to beat your rivals was to translate your ideas into a new piece of hardware or software that everyone liked to use around the office. Making stuff was the most eloquent manner of expression. Pake had always preached the virtues of handcrafting tools in house. Brown said that throughout his time at PARC, "I don't remember buying any technology."[10] Building was the gospel—but it had to be sung in harmony with Pake's corollary measure, Brown recalled: "If you build a tool, use it."[11] Practicing these principles led the tool's inventor and his or her colleagues to be guinea pigs for one another's works in progress. Materially speaking, their workplace—with its meeting areas, individual offices, front lobby, break lounges, cafeteria, and mail and copy rooms—bore some resemblance to many American office environments. And so, if they were supposed to create the office of the future, the logic went, then they would do well to run their experiments on themselves first. Having to rely on the technologies they built would force them to face every day, personally and collectively, the problems and frustrations their prospective users might also encounter. Moreover, as a prototype circulated around PARC's various labs, it gained the chance to attract valuable input from a mixed assortment of world-class technical minds. Perhaps a theoretical physicist would have an oddly useful insight to offer the software engineer, or vice versa, and so on. The potential of each technology stood to be enriched by the spontaneous exchange of disparate expertise.

For Brown, the question of how to cultivate such moments around the building had become a favorite obsession. He would say, without a trace of sarcasm, "The hallways and the coffeepots are as important as the laboratories."[12] These in-between spaces were a breeding ground for offhand remarks that could provoke someone to think differently about whatever problem they had been working on back at their desk. Indeed, it was precisely such encounters—casual run-ins with coworkers far outside Brown's field—that had altered the course of his own research aspirations. They had transformed him, as he would later put it, from a "hard-core computer scientist and an AI junkie" into "a softie, questioning nearly all of the ontological and epistemological assumptions I had embraced in graduate school."[13]

His first projects at PARC, as well as previous research he had conducted at the Cambridge, Massachusetts–based firm BBN Technologies, had revolved around what were then called "expert systems." A major branch of AI, these systems were supposed to emulate the decision-making prowess of human experts. Leading technologists in the late 1960s and 1970s began taking to the notion that they could, for instance, build into a computer program much of the knowledge that doctors drew on when diagnosing a disease.[14] After culling an array of foundational assumptions and nuanced contingencies that populated an expert's brain, the expert system builder would map out an "inference engine"—an intricate chain of reasoning meant to model how experts would likely plot their way from a question to its correct answer, or from a problem to its logical solution. Expert systems represented the most dazzling frontier for many computer scientists at the time. After finishing the workday at BBN, Brown would drive over to MIT ("usually after midnight") to pick up Johan de Kleer, a fellow AI wiz who joined BBN while finishing his doctorate.[15] The pair would roll through the empty streets by Harvard Square and along Boston Harbor, riffing about AI for hours as the city slept. "John always had a thousand ideas," said de Kleer; and de Kleer was a master of the technical side.[16] When Brown left Cambridge for Palo Alto, de Kleer soon followed, and they continued working together on expert systems.

Then, one day at PARC, Brown entered a copy room to chat with one of Xerox's finest repairmen, a guy whose knack for fixing photocopiers had made him a kind of mythical figure in conversations among his peers. Corporate headquarters had tapped Brown to create an expert system that might approximate the repairman's methods in order to guide users through common steps in the troubleshooting process. "I knew the meeting was not going to go well," Brown recalled.[17] The repairman obviously resented the premise of the encounter—that his expertise could be so neatly distilled into a computer program by a few geeks. To make his stance clear, he confronted Brown with a challenge right after they exchanged hellos. "Well, Mr. PhD," the repairman began, "suppose this copier sitting here had an intermittent image quality fault. How would you go about troubleshooting it?"[18] The man mentioned the standard fix outlined in the Xerox manual, which instructed technicians to perform an "image quality test" that required running no less than a thousand copies. Once this twenty-minute copy job was done, the manual prescribed, technicians were supposed to leaf through the pile and find some erroneous pages to hold up against a standard diagnostic sheet. Comparing the latter to the former would reveal the nature of the problem. Brown stood silent, unable to think up a better solution.

The repairman found the manual's advice laughable. He never followed it himself. "Here is what I do," the copy-machine whisperer told Brown. "I walk over to the trashcan sitting here by the copier, tip it upside down, and sort through its contents looking at all copies that have been thrown away. The trashcan is a filter between good copies and bad ones—people keep the good copies and throw the bad ones away. So just go to the trashcan, find the bad copies, and then[,] from scanning all bad ones[,] interpret what connects them all."[19] Brown was stunned. "Brilliant, I thought to myself," he would later write of the interaction.[20] He had noticed the trash can when he entered the room, but it never would've occurred to him to consult it.

Another humbling encounter began with a knock on his office door in 1984. It was a lanky anthropologist this time, named Julian Orr. Like the repairman, Orr was direct: "Nearly everything [you've] written about troubleshooting [is] simply wrong," he told Brown.[21] Orr had also been a photocopier technician before enrolling in graduate school to train as an ethnographer. PARC had recently hired him to do fieldwork on Xerox technicians in action. This kind of research was rare within technology companies, but Brown's old boss, Bert Sutherland, had written ethnographic work into the charter of the portion of PARC that he managed, the Systems Science Lab. Sutherland believed that "basic research would suffocate in a closed environment."[22] Whereas Bob Taylor's Computer Science Lab prized the freedom that came with creating their own little universe, Sutherland's lab played host to a growing roster of social scientists who regularly ventured outside to study how all sorts of people (not just computer scientists) used technology on the job in various worksites. Orr had just returned from months of closely observing copy-machine technicians, interviewing them, and even completing the company's repair training program. By far the most important source of training, Orr told Brown, was the lessons technicians learned during conversations with their peers. Brown's humbling copy-room experience with that expert repairman had positioned him to recognize the truth and implication of Orr's thesis. "Troubleshooting[,] to [the repair technicians]," Brown recalled Orr arguing, "was about making sense of the faulty machine through a story construction process. . . . From this purchase, building sophisticated AI-based job performance aids made little sense."[23] Instead, Orr recommended creating something like a long-distance two-way radio "so each tech rep in a region could easily tap the collective expertise of others in his community."[24]

Brown liked this general idea. Relevant knowledge could be tied electronically into the equipment that workers used at various jobsites; however, unlike an expert-help system, more-fluid technologies inspired by Orr's findings might better accommodate the experiential insights that diverged from the systematic "inference engines" that drove AI-based software. Portable devices like Orr's envisioned radio could also be carried into different contexts and be consulted then and there, rather than back at one's desk. A machine's processing power—its capacity to execute complex programs in rapid succession—was no substitute for the power of knowledge gleaned from paying attention to one's immediate surroundings.

In taking seriously these limitations made plain to him by colleagues outside computer science, Brown started to look for an elusive quality in the technical talent wishing to join PARC. He expressed his outlook on hiring in a piece he sent to the *Harvard Business Review*, which he titled "Letter to a Young Researcher."[25] "There is one qualification we consider more important than technical expertise or intellectual brilliance: intuition," wrote Brown.[26] Intuition was paramount, he explained, because "[PARC's] approach to research is 'radical.' . . . We attempt to pose and answer basic questions that can lead to fundamental breakthroughs."[27] While this sounded cheery at face value, Brown stipulated that it was not. If a balanced and peaceful life was what you wanted, you could find it elsewhere in Silicon Valley—at one of the typical corporate labs where researchers "help[ed] to improve computer technology . . . by going one step farther along a well-plotted path."[28] No such path awaited PARC newbies atop Coyote Hill Road. They would have to find their own footing in the casual, cutthroat, messy system. One week it might serve as a mean-spirited gauntlet, only to feel the next week like an unimaginable carnival of all that was right with the world. You would be freed from the daily grind and made subservient instead to the well-resourced pursuit of your own highest goal. "If you come to work here, you will sacrifice the security of the safe approach," Brown warned prospective PARC recruits.[29] "You will encounter periods of deep uncertainty and frustration when it will seem that your efforts are leading nowhere. . . . But you will have an opportunity to express your personal research 'voice' and to help create a future that would not have existed without you."[30] To a certain kind of technological seeker—exactly the kind Brown hoped to attract—the offer beckoned irresistibly, tugging at the edges of the life they knew.

On Brown's calendar, between upcoming staff meetings, project deadlines, and quick business trips, an hour had been set aside to welcome the latest new recruit. It was a man who'd spent the last two summers at PARC with no intentions of leaving his tenured position on the East Coast. And now, through the gleaming July heat, he was making that long flight west again, this time with his family.

Sitting beside her husband, Mark Weiser, Vicky Reich was still getting used to the idea of California as their plane cruised over the Great Plains and into the American West. They and their two young daughters, Nicole and Corinne, gazed out the window. Neatly plotted farmland suddenly gave way to the Rocky Mountains—golden fields became green forests—followed by vast deserts. The shifting terrain clashed with the character of places they had lived before: quaint, four-season locales like Long Island; Ann Arbor; and, most recently, the Maryland suburbs hugging Washington, DC, where Reich was a librarian at the Library of Congress and had only recently stepped into a new role at the National Agricultural Library. Soon, on account of a bet she'd made with Weiser, Reich would be heading up the Serials Department—magazines, journals, newspapers, and the like—at Stanford.

Earlier that year, she and Weiser had bought a new home in Silver Springs, near Weiser's office at the University of Maryland. Shortly afterward, Weiser received a standing invitation from Brown to come join PARC full time, having done two stints there as a visiting researcher. And shortly after that, Reich had gotten a callback about a job opening in Palo Alto she had not applied for. "There is no saying no to Mark," Reich said. "It's just a waste of energy, so I did what I thought would be a reasonable thing."[31] Fully aware of how badly Weiser wished to join PARC, she told him, without much fear of the fates aligning, that she'd move to California if she had a good job there, too. "Mark, without asking me or telling me," recalled Reich, "started to go through my American Libraries Association magazines and look for jobs, and apparently he wrote my résumé and wrote a cover letter and applied for a job for me at Stanford. The first I heard about this was when they called me up for a phone interview."[32] She did the phone interview, then flew out for an interview on campus, and then began packing up the Maryland house they had just settled into.

Weiser and Reich had been following each other since the day they met as teenagers in 1969, inside the American Museum of Natural History. Each had been invited by mutual friends to come along on a big group trip from Long Island into Manhattan for a day of strolling through the collections. Reich wanted specifically to see the Halls of Gems and Minerals, and only Weiser—an exuberant brown-haired boy she didn't know—volunteered to accompany her. As the two moved among the rubies, the emeralds, and the sapphires, Reich found herself trying to figure out if Weiser had tagged along just to be with her, or if he was actually a fellow geology buff. The more rocks they took in together, the more taken she became. "He was the smartest, most fun person I had ever met, and his eyes were this white-blue. He was so interested and excited about everything. And you could watch his eyes register new ideas."[33] Weiser's way of walking also conveyed a certain energy: his heels barely hit the ground as his toes propelled his body forward and notably upward. This physiological condition, which had stuck since childhood, gave the impression that he was always bouncing, quickly and zealously, headlong into life.

Little about Maryland had pushed Weiser to quit his job as a tenured professor. He was the rising star of the university's computer science department, and even graced its facilities with new computers bought with funds leftover from massive research grants he won early in his career. Promotions happened for him well ahead of the standard timeline, which must've been a source of pride considering that, technically speaking, he had never graduated from college. He made a point of welcoming undergraduates jovially into his classroom, asking them at the beginning of each day's session, "What's good?"[34] Graduate students would confide in him whenever they had "philosophical issues" that needed heartfelt discussing. Weiser was, according to his colleagues, the kind of computer scientist you could really talk with about anything.

But nearly halfway through 1986, a whisper of restlessness crept into his tone. His academic successes over the previous seven years had elevated him to a more administrative role. As the department's associate chair, he was increasingly exchanging memos with bickering faculty about their lab-equipment requests, or appeasing frustrations about their office assignments. Too much of his time suddenly got spent on maintaining diplomatic relations with other units on campus and handling orders from technology suppliers. This new routine of minor headaches seemed to have found an outlet in an impassioned though short-lived hobby Weiser fell into that year: dashing off missives to various businesses about improvements he felt they should make immediately. There was the letter he mailed to Sun Microsystems, maker of the UNIX computer he used, urging them to assemble and ship their user manuals in binders (rather than staple-bound volumes) so that they'd be easier to reference while one typed.[35] Another letter, addressed to United Airlines' "Mileage Plus Policymakers," voiced indignation over the company's stipulation that flights purchased for one's family members could count toward one's frequent flyer account only if said family members shared the same last name as the account holder ("I, my wife, and our children all have different surnames, and this is NOT uncommon among families with two professionals").[36] From the avoidable travesties of shoddy packaging to corporate America's patrilineal fine print, Weiser was quietly lashing out in any small way to change the world. The contrast between his academic sphere of influence and the much wider one he might enjoy in Silicon Valley had grown too pronounced to shrug off. Even just five or ten years of doing his research there—at Xerox PARC in particular—might yield greater impact than a lifetime at his present post. In 1985, moving to PARC was a flattering if unfeasible prospect. By the fall of '86, it loomed like destiny.

In January 1986, a momentous conference took place inside the ballroom of Rickey's Hyatt Palo Alto hotel. The rising generation of Silicon Valley hotshots, in their casual attire, sat circled around tables decorated for fine dining, listening in awe as one personal computing pioneer after another took the stage and told tales of their signature achievements. (Weiser was still teaching in Maryland at the time; he heard about the event later that summer, during his visiting research at PARC. The full-time staff members who had attended the conference were still talking about it then, and Weiser hung on every detail they recounted.)

The January gathering had been billed as the "ACM Conference on the History of Personal Workstations," but that title belied its gravitas. It was not a routine scholarly proceeding. Ranging widely in age, the fourteen men who spoke at the ballroom could have all been rightly called founding fathers of the digital revolution. The elders went first. There was Gordon Bell, chief architect of landmark 1960s "minicomputers," including the Digital Equipment Corporation's PDP-4 and PDP-6—precursors to the PDP-7 Weiser used almost daily throughout his twenties. There was Larry Roberts, who got introduced to the audience as "the father of packet switching" for his seminal contributions to the ARPANET, the network that initially linked computers at select American research universities, where Weiser and many of his contemporaries had their first online experiences. And there was Douglas Engelbart and J. C. R. Licklider.

Of the men who spoke at Rickey's Hyatt, Engelbart and Licklider most exemplified the nebulous role that Weiser hoped to grow his way into at PARC. Though none of the research Weiser had done at Maryland could even generously be called visionary, he had long considered himself a thinker above all else. Engelbart and Licklider were both technological thinkers; it was the concepts they built into their essays and lectures during the '60s and '70s, more than the technical advances they shepherded in the lab, that defined their legacies. Their respective ideas both guided and transcended the early prototypes they had worked on, as they each put forth a broad framework for imagining how connected, interactive computer screens might come to aid knowledge workers across a range of cognitive tasks.

Engelbart had come to computers not out of a technical infatuation with gadgetry, but rather from a hunch that human-scale electronic displays could provide a problem-solving environment uniquely suited to the growing complexity of a highly specialized and quickly globalizing world. In 1945, when Engelbart was a young navy radar technician stationed in the Philippines, Franklin D. Roosevelt's chief science advisor, Vannevar Bush, warned the nation that "a growing mountain of research" promised to overwhelm any individual scientist's processing ability.[37] The paper-based forms of communication that enabled the Enlightenment's Republic of Letters, Bush argued, could no longer keep pace with the speed and scale of informational demands arising from the Second World War. Reading Bush's popular essay, "As We May Think," in the Philippines set the agenda for Engelbart's research career. With a sliver of funding from the US Air Force, Engelbart went on to establish a lab at the Stanford Research Institute during the mid-'60s, where he and his team devoted themselves to reimagining computers as personal workstations that could "augment the human intellect."[38] Their favored tactics for doing so differed crucially from those Bush had put forth. Whereas Bush's envisioned "memex" system drew on 1930s microfilm technology in hopes of giving individuals easy and rapid access to unprecedented masses of documents, Engelbart wanted to create an electronic workspace that groups of researchers could share. In addition to exchanging documents over their screens, Engelbart's imagined users would collaborate instantaneously over great distances. Just as writing upon clay tablets, parchment, and paper augmented one's ability to retain and analyze stored information, a network of personal electronic workstations would, Engelbart believed, greatly expand the base of readily available knowledge, which dispersed working groups could jointly examine, modify, and reflect on as they endeavored to address multifaceted issues.

Engelbart's efforts in Palo Alto had been augmented crucially on two fronts by Licklider's work on the East Coast. Engelbart's first injection of major support came from Defense Department funds that Licklider presided over inside the federal government's Advanced Research Projects Agency. The money enabled Engelbart's team to start building their suite of devices, which promised to accomplish the goals Licklider himself had set for computers a few years earlier in his 1960 article "Man-Computer Symbiosis." Licklider, who had entered computing by way of a doctorate in psychology, insisted that computers would become more widely useful if people could enlist their powers *while* they thought about a problem, instead of just feeding the computer a problem they formulated beforehand on punch cards. Involving the computer as an intimate partner in science and creativity entailed major design challenges, which Licklider enumerated in detail. Foremost, the computer's inner workings needed to be translated into a language that users could recognize; in turn, users should be able to manipulate the data by means of gestures and symbols closely related to the ones they employed when communicating with other humans; requiring people to look up obscure codes every time they needed to perform any function stopped far short of symbiosis. Then there was the matter of inputs and outputs: Which kind of display equipment and control mechanisms would best facilitate "real-time thinking" between a human and a computer? While Engelbart's lab was not the only group inventing answers to Licklider's questions, the Stanford-based team

A student at Brown University using a light pen to select text on an IBM 2250 display in 1969. Photograph by Gregory Lloyd.

had by 1968 caught the attention of technologists everywhere with their prototype—the "On-Line System," which came to be called NLS.

The NLS famously introduced the computer mouse, a new input device that had been made operative by Engelbart's top engineer Bill English. As the natural weight of Engelbart's palm rolled this small wooden block over the surface of his desk, a little black dot moved around on his cathode-ray-tube display, allowing him to select and modify the electronic texts on screen without having to type a coded command. Anyone who had during the 1960s used a "light pen"—a short-lived, cumbersome mechanism that IBM users rubbed against their screens to highlight text—could instantly appreciate the mouse's easy speed, comfort, and versatility.

For text entry, the NLS featured two additional input devices: a QWERTY keyboard and a less familiar keyboard consisting of five piano-like keys that could be struck in various combinations with one's left hand to produce each letter of the alphabet, leaving one's right hand free to remain on the mouse. These three inputs, coupled with

Douglas Engelbart using the NLS, featuring the three-button mouse (at right), QWERTY keyboard (at center), and five-button "chord keyset" (at left), circa 1968. Courtesy of the Smithsonian Institution.

the crisp projection of working documents on its screen, enabled NLS users to create and edit electronic files with a sense of immediacy that rivaled pen and paper. You could point, click, and type just about as quick as you could think. Once you input text into Engelbart's system, you could then do things with the words that no typewriter or printed book would ever let you do. The NLS, according to the historian Fred Turner, "offered its users the ability to work on a document simultaneously from multiple sites, to connect bits of text via hyperlinks, to jump from one point to another in a text, and to develop indexes of key words that could be searched."[39] Unfamiliar functionality of this sort was the keystone of a collaboration-centered foundation Engelbart wished to establish for future computers. The individual workstations through which his team accessed the NLS were personal in appearance only. Each terminal was principally a gateway to "a shared intellectual space" where multiple users could follow one another's minds as their respective mouse pointers moved across the same document.[40]

History had already memorialized a presentation in December 1968 during which Engelbart made a show of the NLS's novel features for enthusiasts at the ACM/IEEE-CS Fall Joint Computer Conference in San Francisco—a presentation that the tech journalist Steven Levy later termed "the mother of all demos."[41] From the stage at Rickey's Hyatt, eighteen years later, Engelbart could still bring computer scientists to the edge of their seats whenever he turned to share film clips of that old demo. Encapsulated in his footage were the broad outlines of what personal computing had come to look like in the '80s, as well as some finer details the industry hadn't found a way to incorporate. Many attendees in the crowd at the January 1986 commemorative gathering (including the speakers set to take the stage next) remembered how mesmerized they felt the first time they saw Engelbart demonstrate the NLS. Some parts of that vision had been realized, and other parts seemed further out of reach now than they had then.

After Engelbart's presentation came presentations by three former Xerox PARC researchers: Alan Kay, Butler Lampson, and Charles Thacker, the principal designers of the Xerox Alto, the 1973 machine that secured PARC's place in technology history. The Alto, heavily inspired by aspects of Engelbart's NLS, set the course of personal computing more surely than any system that preceded it. As the author M. Mitchell Waldrop put it, "The Alto was certainly not the first personal computer . . . but [it] was the first machine that most of us would recognize as a personal computer."[42] The Alto departed from the then-ascendant model known as time-sharing, championed by Licklider and Engelbart, in which users sat at multiple terminals and each took turns tapping into one large mainframe. The Alto did not rely on a mainframe; PARC computer scientists stocked the Alto with enough onboard memory and processing capability to support a single user on its own power. Another young PARC researcher, Bob Metcalfe, promptly spearheaded "Ethernet," a network over which numerous Alto machines running in the building could share programs and files. The fact that Altos could exchange data without having to divvy up any of the hardware, as mandated in time-sharing systems, left each machine with the resources necessary to support high-quality graphics on a bitmapped display. Under the bitmapping approach, explained Thacker, "each screen pixel was represented by a bit of main storage."[43] Each pixel could thus be precisely controlled and made to serve whatever software a user ran. Manipulating individual pixels in this manner made the Alto's screen immensely more versatile than earlier displays, which generally presented text in only one font. In the windows of the Alto's word processor, called Bravo, the pixels formed letters and punctuation marks over a paper-like backdrop. In Markup, the Alto's drawing program, the pixels would turn into squiggly lines and shapes as users moved their cursors over the screen to create pictures. The pixels were also molded into icons and menus, each of which could be clicked to perform a corresponding operation, such as opening a folder or saving a file—sparing users from typing codes into the command line. In total, the convergence of these innovations amounted to a new kind of computing experience that came to be celebrated as the "graphical user interface." Popularized in 1984 by Apple and later Microsoft, the graphical user interface, or GUI, was by 1987 well on its way to becoming the industry standard.

Now that Weiser was joining PARC, events like that January 1986 conference might be a notable part of his regular schedule. He would travel in the same circles as Engelbart and the others, perhaps even bump into them at the grocery store. The projects he'd work on stood a chance to initiate some next big phase in computing. When ideas caught on in Silicon Valley, they tended to spread far and wide. As those fourteen researchers told their stories, their innovations were woven together into a grand narrative, putting on full display the enormous impact that a hunch could grow to have when nurtured in the right lab.

The event demonstrated something else, too, which Weiser probably found even more striking. A few of those pioneering researchers who spoke at the January conference had alternative prospects on their minds. Even as they were called to wax nostalgically on the personal workstation and the legacy of early desktop systems like the NLS and the Alto, they peppered their remarks with mentions of prototypes-in-process that might someday challenge the industry's recent uptake of the mouse and keyboard, the stationary monitor, and the graphical user interface.

Licklider mentioned with great interest "the idea of instrumenting the body of the user."[44] He said the physical keyboard was "clearly on the way out" and wagered that it would be replaced by "instrumented fingers"—future users could wear thin fingernail attachments, allowing them to select content on a screen by touching it.[45] He believed that people would speak to computers and that the machine would be able to speak back. Engelbart had showcased on his slides a lesser-known configuration of his lab's technology that wasn't seen in the heralded 1968 demo. He talked the ballroom audience through photos depicting a research-staff meeting in his lab's conference room. His staff sat at tables that formed a rectangle; in the middle were several computer terminals, each positioned lower than the tables and tilted back so that each man needed to peer down at it. This unusual setup, Engelbart explained, allowed the staff to focus primarily on one another and secondarily on their connected screens. In this case, their engineering effort reached beyond hardware and software; they took care to situate the terminals around a particular group activity and tried to make computing serve that activity.

Then, of course, there was Alan Kay's Dynabook vision—a resolutely mobile concept that had been the focus of his talk. Kay had been thinking about the Dynabook prior to joining PARC in 1971, and had continued pursuing the idea since his departure in 1981. Before the Xerox Alto got its name, Kay and colleagues initially called it the "Interim Dynabook." The Alto was, in essence, the best kind of personal computing they could actually build using early '70s technology. It was just a step, albeit a giant one, toward Kay's dream of "the computer that goes with you."[46] As Kay showed the crowd his diagrams for a tablet-like device—an early iteration of his Dynabook idea—he emphasized that the Dynabook was not reducible to one form factor. "It is not a box," Kay said. "What it is is a service. . . . It's a Dynabook if it gives you your information services wherever you are on earth."[47] Mid-'80s computer science had still not reached a point where small, portable devices were very capable in this respect, but Kay remained confident that electronic tablets would someday make for a better, more personal interface than any desktop machine.

One of Kay's collaborators and a codeveloper of the Smalltalk programming language, Adele Goldberg, edited the January conference proceedings for publication. She concluded her introduction to the volume by underscoring how so much of the thinking that led to personal computing ultimately pointed beyond it: "One gets the feeling that the personal workstation was not the goal at all, and that [its] history is about some as yet unknown artifact or alternative to the way things are."[48] While the CEOs of Silicon Valley were banking on a future defined by desktops and laptops, these researchers whose breakthroughs inspired those products were already keen on other developments.

As Weiser would soon reveal to his new colleagues at PARC—the veritable birthplace of personal computers—he, too, was eager to push computing in a different direction.

On his first day at PARC, Weiser opened the door to his new office and sat down to the computer waiting on his desk. Getting it set up was the first order of business. He had set up many before. There was his PDP-7 minicomputer that, as Reich recalled, "took up the whole damn living room" of a cramped Ann Arbor house she and Weiser had shared with a few friends during their grad school years.[49] Then, in Maryland, on his Xerox Star and his Sun-2/50 desktop workstations, Weiser developed software for his research and created games for his two little girls. At home, Corinne and Nicole had watched in awe as he assembled an Intel-based Heathkit microcomputer with its matching terminal and floppy-disk drive. They would grow up to retain many fond memories of the way their father brought new technologies home to bring the family together. Whether they were programming the little robots he'd give as holiday gifts or playing the custom versions of Tetris he made, computing was throughout their childhood an exciting group activity.[50]

They often played music around the home—Weiser loved the drums—and the times they gathered around a computer had a similar feeling.[51] Hitting notes over a shared beat and pressing buttons on a shared device were both fun ways to create and think in unison. At the same time, Weiser also tried to connect with distant strangers from home. His had been among the few households in Maryland to regularly log on to the Usenet, one of the earliest publicly accessible online communities where anyone with the right equipment could participate in the network's user-managed forums. Weiser threw himself into the Usenet experiment and made a hobby out of posting casual reflections about his daily life to various message boards where other Usenet enthusiasts offered up their thoughts. The applications Weiser gravitated toward in his free time always meant to make the computer into something of a shared intellectual space, in the tradition of Engelbart's vision.

The many hours Weiser spent on computers at work were far more solitary. As a young software engineer, he enjoyed the puzzle-solving moments that typically riddled each new programming job. As a scholar of computation, he was drawn to the challenges that arose during attempts to connect different technologies into a cohesive system. But the thrill of crafting order from complexity was usually a hard-won pleasure, and his work invariability kept him sitting alone at a screen in a university lab. "One of the disadvantages of being involved with Mark Weiser," Reich said, "was that he was gone a lot, in a computer room."[52] And while Weiser found his technology research rewarding and worthy of such immersion, there were elements in his long-standing relationship with various computing systems—from batch processing to time-sharing terminals and PCs—that he loathed. Computers were always interrupting the rest of his life. He had to take time out from teenage dates to go swap out decks of punch cards on campus during his allotted time slots. He'd often have to stand in line to use a time-sharing terminal for longer than it took him to complete his task once he finally got access to an open machine. "Just the bureaucratic nonsense of getting the computers to do the work that he wanted them to do for him," said Reich—"it was a frustration, constantly."[53]

Kay's proposed Dynabook ("the computer that goes with you") held obvious appeal for Weiser. The Dynabook idea, more than the Alto prototype, was the PARC innovation that resonated most with his sense of what computers ought to become. Of course, shrinking the PC on his desk down to the size of a tablet would not in itself transform computing into a more collaborative activity performed with others in a shared space. But making powerful devices that were also portable was certainly a start. A handheld wireless tablet called to mind a future in which people could tap into the power of computation from all sorts of locations. One would not have to keep putting life on hold in order to perform electronic duties at a campus mainframe, a terminal in a time-sharing lab, or a desktop in the office.

Weiser eyed the clock and saw it was time for his welcome meeting with Brown.

Brown waved Weiser into his big office. There were all the usual topics to discuss with the newly hired: how was the move, which colleagues had he met so far, what projects would he like to initiate, and so on. None of that appeared to resonate with Weiser's present agenda. "He had this kind of creative sparkle in his eyes all the time," recalled Brown.[54] Weiser was itching to discuss something else.

"You're going to find this stuff strange, John," Weiser began.[55] He then launched into a fifteen-minute whirlwind tour of European philosophy, with frequent allusions to a controversial thinker named Martin Heidegger. Weiser had spent his first year in college reading and rereading one of Heidegger's books. The ideas he had absorbed from those pages still gripped his imagination, especially the philosopher's theory of entanglement—this wonderful notion that no one was ever alone. Or, phrased a bit more cerebrally, it postulated that human existence—our actions, perceptions, even our thoughts—never occurred in a vacuum, no matter how often it felt that way. Being connected to the world and to others was our natural, indelible condition. Isolation was an optical illusion under Heidegger's framework, albeit an illusion Weiser had known well.

Weiser then insisted, more to the point, that these ideas and more like them offered a radical new basis for redefining what computers could be. He wanted, somehow, to invent digital technologies that fostered in people this sense of entanglement, but he didn't know yet what that would look like. But it had to look different than an individual sitting and staring at a screen. After a while, Brown started to laugh.

"You call that strange?" Brown interjected. "That's what we've been talking about already."[56] He mentioned PARC's anthropologists. They were steeped in similar ways of thinking and had made some headway on showing their technical counterparts the merits of emerging worldviews that stressed embodiment, relationality, and context. One anthropologist, Lucy Suchman, had read Heidegger as well, and she was busy advancing her own critique of the ways computer science operated at a remove from life as people tended to experience it. It was enticing to muse about what she and Weiser might work on together.

In any case, Brown was thrilled at Weiser's readiness to digress passionately into seemingly far-flung subjects that few other R&D outfits would take seriously. "I was just blown away," Brown remembered. "Here was this computer scientist, and he wanted to interview me on philosophy."[57]

Once Weiser left, Brown said to himself, *He's going to fit in perfectly here.*[58] Weiser certainly hoped he would fit in, too. But he was also, if only in the notes he wrote to himself, still in the habit of doubting his ability to connect, fully and truly, with anyone.

2

The Innovator as a Young Seeker

AS A TEENAGER, Mark Weiser had a saying that often startled people: "The computer [is] my best friend."[1] He'd say it to the bemusement of teachers or neighbors whenever they probed him about his unusual pastime. He'd say it with a smile, his blue eyes beaming through his glasses, his shaggy brown hair slumping to the side. While the line may have garnered sympathetic laughter in the moment, there was some truth to it.

It was 1968, and Mark was one of the few sixteen-year-olds who knew much about computers in Stony Brook, a cozy town on Long Island's marshy North Shore. He had taught himself programming a year earlier and was soon recruited into a summer job writing computer programs at Stony Brook University. The gig had required Mark to move to Stony Brook by himself. His parents and two sisters were still living upstate near Ithaca, where his father held a temporary post teaching chemistry at Cornell. In the campus basement that housed Stony Brook's most coveted technology, Mark's infatuation with computing blossomed in the long silences of a season alone. Reflecting on that summer years later, Mark wrote, "The computer became what kept me going. A successful program was in many ways what I lived for. . . . Just using the fancy glittering device, probing its intricacies . . . I was looking for something to fill my life, grabbed at what was there, and became a slave to it."[2] The rest of his family relocated to Stony Brook that fall when an enrollment surge allowed the growing state school to offer Mark's father a full-time professorship. Mark insisted on continuing his programming job when the new school year started; his evenings and weekends revolved around punch cards.

The devices Mark worked on then would appear neither fancy nor glittering to present-day onlookers. They featured no images, no icons—not even a mouse or screen. Visiting his electronic best friend required Mark to inhabit the sterile, heavily air-conditioned bowels of the Stony Brook University Computing Center. Like most academic computing hubs of the era, the center occupied a series of fluorescent-lit, linoleum-tiled rooms filled with loud machines and stacks of intricately designed paper. The mainframe computer—there was only one—sat in a room that was walled off to everyone except the small staff tasked with operating it. Mark could catch only glimpses of the mainframe through two windows: one where he dropped off his punch cards for processing and another where he picked them up afterward. He and the other programmers each did their computing work at some distance from the mainframe, in another chilly room lined with card punch machines like the popular IBM 29.

Seated at the card punch machines, campus programmers would enter their data and type out commands on the keyboard, which would with each keystroke tell the machine to poke tiny, precise holes into a punch card; the pattern of these holes on each card was what the mainframe understood. Before a deck of cards was submitted to the operator for batch processing, it was scrutinized by the programmer for errors. Even the smallest mistake on just one card could result in a botched job. If that happened, you would have to take the erroneous card back to the card punch machine, fix it, and then sign up for the next available time slot to drop off your deck for a second run.

As demanding and unforgiving as this process was, Mark found programming enthralling. He was and would always be a lover of puzzles. Figuring out the best way to tell the computer exactly how to compile, calculate, and organize the data you gave it was a never-ending puzzle. Once your punched cards aligned to form a good program, the computer could produce simulations and predictions of unparalleled scope and complexity. Scientists, engineers, governments, and cor-

An IBM 29 card punch machine. Photograph by Joe Mabel.

porations used IBM's FORTRAN language to craft large-scale models of physical, social, and financial systems. Mark lamented having to structure his teenage life around the mainframe's tight schedule, but few things lured him to miss his assigned time slots at the computer center.[3] Though his punched cards were his only means of communicating with the computer, he still found it to be a dazzling, empowering companion. Writing programs at the card punch machine did provide users with the sense of a high-tech space where they could sculpt their thoughts into intricate, dynamic systems. Mark could pass whole days absorbed in lines of hole-punched code the way architects inhabit buildings-to-be in their blueprints.

While Stony Brook had its Computing Center, the small college town was no Silicon Valley. Brookhaven National Laboratory, a hub for scientific and technical research geared heavily toward atomic energy, was located twenty miles east; but it was remote from Mark's high school life. Mark had come into computers mainly through his father. David Weiser, by virtue of his various university posts, became, to put it mildly, an early user of card punch machines (and, later, time-sharing systems). The chemistry professor may have been one of the world's first computer addicts, though that was the least of his vices—his diary entries detailed a masochistic drive to remain at the desk long after his productivity waned, in between steady helpings of alcohol.[4] When David finally did come home, he usually struggled to fall asleep. Exhausted and overwhelmed, always thinking and often intoxicated, he was distant and withdrawn even when he was present around the house.

David probably would have been fired many times over if not for the efforts of Mark's mother, Audra Weiser. In her diary she noted how she would labor to get David out of bed and make him presentable enough to teach his afternoon classes. Night after night, she phoned him on campus every half hour, often starting at 11 p.m. and sometimes calling until 1 a.m., urging him to stop computing, come home, and get some rest so he could begin the next day on time. David took tranquilizer pills to give himself a chance at sleeping. If the pills didn't work, Audra would awake to find him in the guest bedroom, drinking and working. On one such occasion, as Audra entered the room to fetch him, David shushed her away, saying he was "having a peak experience."[5] That alcohol-fueled state must have seemed a reprieve from the rejection notices, missed deadlines, and unmet expectations that stained his research career.

In Mark's youthful eyes, his dad was a dedicated scientist and the computer was his medium for discovering new knowledge. This impression was justifiable. David had begun his career at the University of Chicago under the wing of a Nobel-laureate chemist, and he still exuded brilliance in spite of his shortcomings. Though David achieved little as a scholar, his propensity for solving the most confounding mathematical puzzles earned him high praise in unexpected places—the popular mathematician and *Scientific American* columnist Martin Gardner, who was famous for the math problems he challenged his readers to solve, once singled out the work David sent in, calling it "one of the clearest analyses I received of a problem."[6]

The son imitated the father dutifully, regarding himself as a researcher from an early age. For both, "doing science" was too important to let anything else get in the way. Consciously adopting "the social awareness of a pure scientist," the young Weiser declared: "To find truth is an adequate end to justify most any means."[7] With David pecking away at a keyboard on campus, Mark's solitary hours working at a keyboard of his own may have been, along with books, his best means to feel connected with his father. David's love for his children surfaced in the affectionate gazes he cast at them during family photographs, but his feelings were much harder to detect when a camera wasn't around. Those who visited the family home rarely observed David exhibiting any behavior that might be described as "interactive," let alone fatherly.[8] He did, however, leave a trail of academic monographs and scientific journals scattered about the kitchen and living room, like breadcrumbs of the mind. Mark would pick them up in David's absence to read what he had been reading.[9]

Their implicit bond was held together by Mark's uptake of the objects that kept David's attention. What his father had focused on, Mark would focus on as well, separately. Audra once tried to intervene, confiding to David and Mark that their long working routines, their hours gone computing, had pushed her to the brink of depression.[10] They both agreed to return home every night for dinner. David's commitment to family dinners soon wavered, then dissolved; but during those few weeks, "life was lovely for a while," Audra noted in her diary.[11]

In addition to computers, David—or, the texts David left strewn everywhere—introduced his son to philosophy. Paging through David's piles led Mark to grapple with concepts that altered him in ways that proved formative and lasting. Mark would later return to them constantly for inspiration and guidance throughout his career at Xerox PARC. In David's failures lay the seeds of Mark's future ambitions. David's inability to produce much publishable research as a chemist led him to teach and read widely in the philosophy of science, a nascent cross-disciplinary field that brought the ideas of Descartes, Hegel, Kant, and others into conversation with the theories and experiments of modern physicists, chemists, and biologists. David's syllabus for "Chemistry 286: Basic Principles of Quantum Mechanics" began with an epigraph from the novelist Franz Kafka and proceeded to synthesize Galileo's insights on particle acceleration with Nietzsche's remarks about the tenuous relationship between nature's reality and humanity's ability to perceive it.[12] His students at Stony Brook were fascinated by David's knack for drawing provocative connections among disparate texts, and he managed to be a well-liked teacher.

Mark fancied himself as his father's most promising young pupil. One of David's favorite volumes, *The Tacit Dimension* (1966) by the Hungarian philosopher Michael Polanyi, became Mark's favorite, too. Mark devoured it with an exuberance that left the volume shrouded in boyish wonder for the rest of his life (he would later name his Silicon Valley startup after Polanyi's book). *The Tacit Dimension* presented a whole new theory of knowledge in less than ninety pages. And while its author famously coined the term *tacit knowledge* (which remains a popular descriptor for learning that cannot be formally taught), his theory went deeper than that. Polanyi argued that human experience was more or less tacit by nature; all of us, at every moment, "know more than we can tell," he wrote.[13] Our relationship with the world around us always includes more sensory information than our conscious minds can process. According to Polanyi, this omnipresent layer of peripheral impressions is not merely white noise. All formal knowledge is rooted in, born out of, and filtered through this tacit dimension that both evades our understanding and makes understanding possible.

For example, Polanyi asked his readers to consider our ability to recognize the faces of people we've met before. Save for the visually impaired, most humans possess the capacity to recognize thousands of different faces without conscientious study. And yet, even as we might easily spot a friend's face in a large crowd, we would be hard pressed to recite from memory the eye color associated with each of the many faces we recognize at a glance. Our explicit knowledge of our friend's eye color, let alone the subtle contours of their nose or jawline, pales in comparison to our tacit sense of how the individual features look as a whole. We couldn't describe every facial detail, but we know with absolute certainty whether or not the person seated across from us is indeed our friend, or our mother, our boss, a stranger. We know much more than we can tell.[14]

After an illustrious career as a chemist, Polanyi had taken up philosophy to contemplate how scientists thought their way through the research process. He was particularly curious about intuition—moments when scientists experience a kind of involuntary mental itch that they felt compelled to investigate in the absence of an established scientific reason for doing so. In Polanyi's estimation and in his experience, all breakthroughs in knowledge began with a tacit hunch. Following a hunch meant taking a brief leave from objectivity, precision, and other cardinal values that science is known for. Using a range from lab experiments to artistic works and medical diagnoses to mundane tasks, Polanyi brought his varied examples together in support of an overarching dictum: the scope of our knowledge extended well beyond our consciousness; when we valorized only that knowledge which we could codify and transmit clearly, we diminished our creativity, the tacit genius of intuition, and the inherent richness with which we might otherwise perceive the world.

Polanyi was just the sort of writer capable of captivating a smart kid who moved restlessly from one grand notion to the next. When you really thought about it, as Mark did, conceptions of knowledge were at the heart of science and technology. And the scientific community's traditional reverence for positivism—an epistemological stance that rejects all claims that cannot be stated with empirical certainty—lent little credence to anything "tacit." Working from an alternative theory of knowledge had enormous implications for how one might conduct scientific research or pursue technological innovation. Science was, in essence, the search for testable, reproducible knowledge; and technology, in its many modern varieties, was usually a means for sharing that knowledge.

The whole scene of mainframe computing, it seemed to Mark, was particularly removed from the tacit dimension. First off, the fact that you had to descend into the underworldly depths of a computing center every time you wanted to compute anything meant that you were habitually having to limit your sensory experience to what was on offer there. Using a computer or a card punch machine meant isolating yourself from the sights, sounds, and smells that Polanyi deemed crucial to intuition. In the computing center, you were confined by the needs of machines that had to be walled off from the elements and kept secure from curious, untrained passersby. Both materially and socially, computing needed to occur in a separate context built around the computer's best interests. And then, once you were down there, there was the act of programming—the days and nights spent gazing at little else but punch cards moving across the card punch machine, and at the explicit knowledge the cards displayed. Crafting a functional program out of code, as Mark had spent so many hours doing, involved reflecting on all the cognitive tasks that comprised a routine mental activity and outlining the sequence of each step in a series of discrete commands the mainframe could follow. It was about distilling and extracting the logical syllogisms of human thought from the tacit context of our embodied lives. From the isolating place you had to sit in to the abstract headspace you had to cultivate while coding, the time spent inside a mainframe computing center revolved around the production of explicit knowledge by means of explicit knowledge. A version of this sentiment would surface again in 1991, when Mark would insist that, even in the case of desktop PCs, "The computer nonetheless remains largely in a world of its own."[15]

Reading Polanyi's celebrations of tacit knowledge cast a different light onto the computer simulations that captivated Mark. Suddenly, computers as Mark knew them felt less like a new frontier and more like some far-off planet. Polanyi had declared, "It is impossible to account for the nature and justification of knowledge by a series of strictly explicit operations."[16] The critique was not directed at computers, but it did apply more exactly to computer programming than any other domain. Almost as soon as Mark had fallen in love with code, he began to mull over the idea that something essential got lost in translation whenever you attempted to model the world on a computer.

As he read further into Polanyi, however, Mark took notice of passages that underscored how some devices accentuated our connection to the tacit dimension of everyday lives. One of Polanyi's examples would continue to captivate Mark through the 1990s, as he incorporated the account into an essay of his own that catapulted him to the forefront of the tech industry. Polanyi described the scenario of a blind person using a white cane to navigate a busy sidewalk, to showcase how tacit knowledge could be facilitated by tools. When the visually impaired tapped their canes against the ground as they walked, they gathered an extra layer of sensory detail about the present conditions in the environments they traversed. Commenting on his own attempts to use a white cane, Polanyi wrote: "As we learn to use a probe, or to use a stick for feeling our way, our awareness of its impact on our hand is transformed into a sense of its point touching the objects we are exploring."[17] White-cane users attended *from* the cane's vibrations *to the sidewalk's* contents. This *from-to* dynamic, exemplified by the cane, constituted for Polanyi "the basic structure of tacit knowing."[18]

The white cane was an information processing device, in its way, and it operated very differently than computers were operating, from mainframes to the time-sharing systems coming online at select universities during the late 1960s. Whereas computer users typically attended *to* the machine's printouts or an electronic display in order to engage with data pulled *from* various places, tacit technologies like the white cane served as conduits affording their users a new access point for processing information emanating from their surroundings. Computers were unparalleled reservoirs for storing codified data, which one could access via codified instructions. But if technologists wanted to boost humanity's capacity for tacit knowledge, then they had much to learn from the humble white cane. The white cane delivered new information to the hands of its holder by treating the ground as an input that it converted into an output of tactile vibrations. It was a mechanism for transcoding data about the physicality of various objects into a meaningful, bodily sensation that could be perceived and readily understood by those who could not see the objects. Where the computer tended to diminish one's contact with tacit stimuli, the white cane brought new information to people in a manner that reinforced and even extended their sensory grasp. When considered from Polanyi's perspective on knowledge, this scenario of a blind person using a white cane to navigate a busy sidewalk illustrated technology at its best. In Mark's thirties, this scenario become a kind of a North Star orienting his future innovations.

But for a while, Mark wondered if computers were simply incompatible with the tacit dimension. For the next couple years, he temporarily strayed from programming to gain a greater hold on this new sort of life that Polanyi's philosophy spurred in him.

Mark finished his last year of high school with his mind trained 1,200 miles to the south. He and Vicky Reich had begun dating, and they were both open to a long-distance relationship as she stayed in Stony Brook for her senior year. An experimental college of some five hundred students had just graduated its first cohort, and word of its eccentricities spread quickly up the Eastern Seaboard, alluring bright kids who were seeking something different. Mark's university applications boasted all the hallmarks of an impressive seventeen-year-old: National Honor Society, great grades and test scores, Advanced Placement courses, and some very high-tech work experience. He was a natural fit for the Ivy League on the merits of his record, or for a school like the University of Chicago, his father's alma mater, not far from where Mark himself had been born. Off the record, though, he came from an all-but-broken family. and he wasn't sure what he wanted to do. His desire to get away from it all stood at odds with his accomplishments and the expectations they raised. This tiny new college in Sarasota, Florida, seemed like a beacon for a fresh start.

Opening in 1964, New College of Florida had abandoned nearly every academic norm that still orients universities today. There were no grades; professors wrote narrative evaluations for each student; there was no required curriculum. Orientation materials made this clear to first-year students: "You [are] in the position of creating and designing your own education from the start."[19] A campus-wide attendance policy deemed every absence excusable. The only rules limiting behavior within the utopian confines of this adolescent republic were a precious few agreed upon by the student executive committee. Most schools had homecoming celebrations; New College threw an annual campus-wide party called "PCP," named for the hallucinogenic drug phencyclidine. In the spring, graduating students walked on stage to accept their diplomas wearing any costume they wished (the birthday suit soon became a tradition upheld by several daring students each year). Measured in terms of flower power, illicit substances, and progressive idealism, this grassy patch of land on Florida's sleepy Gulf Coast could have passed for several blocks in Haight-Ashbury. At the same time, by 1969, New College's acceptance rate and the caliber of its incoming students held pace with Yale and Princeton—much to the delight of its founders, who had ordained their campus by mixing a shovelful of Harvard dirt into the local soil.[20]

Mark arrived on campus in the fall of 1970 and was immediately confronted with the question of what fields to study. Not what major he would choose—New College students weren't encouraged to think like that—but rather: What intellectual predicament or social problem was he compelled to solve? Spared from the child's play of grades and quizzes, which giant's shoulders would he stand upon? An elite, do-it-yourself education instilled its own brand of imperatives and pressures. Freedom to find their own way across all disciplines unwittingly encouraged students to stick with what they already knew. Mark was determined to search for answers to the questions Polanyi's book had filled his mind with since eleventh grade. How did people come to amass knowledge about the world, and what concepts would help illuminate the mysteries at play between us and the objects we encountered? Which technologies attuned us to the richness of our surroundings and which ones got in the way?

Mark split most of his days at New College wandering among the palm trees to the campus library and back to the typewriter in his dorm room. Absorbed at his desk, he wrote poems and short stories every week that functioned as a literary analog to the philosophical texts he consumed at the library. The library was small but spectacular. Housed in the bayside former estate of the circus baron Charles Ringling, this Gatsbyesque mansion gifted its patrons an uncanny pairing of free books and ocean views, with the entire spectrum of blue glittering up from the sea to the sky. Taking in this scene at a second-story window behind a pile of worn hardcovers was, for Mark, the definitive New College experience. Evenings lounging in vinyl chairs with a few friends around the campus pool rounded out his routine. At the pool, as twilight insects buzzed and chirped amid the humid darkness, Mark learned about the Chinese board game go. He played and watched others play, night after night, pleasantly absorbed by the game's complexity.

There were classes and chats with professors some days, too, but Mark did not build his schedule around them. He attended when the mood struck or whenever he needed a break. The conversations he found most pressing were the ones he was having with books.[21] He was chasing after a new way of feeling at home in the world—seeking a way out of his own head—and he pursued it largely in seclusion.

At New College, Mark discovered a new guide who might take him further than Polanyi had. Mark's freshman year revolved around reading Martin Heidegger, an enthralling and strange German philosopher whose 1927 masterpiece, *Being and Time*, was named the second most influential book of the twentieth century in a survey of North American philosophers.[22] The same thing that had drawn Mark to *The Tacit Dimension* again drew him to *Being and Time*. Like Polanyi, Heidegger placed great emphasis on the simplest of objects. Both thinkers regarded tools as the linchpin of humanity's most creative and authentic forms of living. Polanyi valorized the white cane; Heidegger revered hammers. Their mutual love of hand tools sprang from a shared, vague worry that modern societies dwelt too much in vapid abstractions and that, as a result, people were losing their intuition. They feared the mind was growing distant from the body on a cultural scale. In Heidegger's analysis, he issued a sweeping diagnosis, tracing the problem back to an original sin in ancient philosophy, which had since been fatefully compounded by René Descartes.

Descartes had loomed over Mark's early forays into philosophy because Descartes's worldview mirrored aspects of his teenage experience. "The Cartesian idea of man is intensely introspective," Mark wrote in a high school term paper that bore the droll tone of a writer who knows his subject all too well.[23] Descartes's notion that the self-talk of our interior monologues constituted the ultimate bedrock of reality should have comforted Mark. He was coming of age amid pop psychology's self-help heyday. With war marching on in Vietnam and protesters in the streets, broad swaths of the American reading public had turned inward and propelled books like *I'm OK—You're OK*, *The Magic of Thinking Big*, and *Psycho-Cybernetics* onto bestseller lists. It was in line with this trend to silently ask and answer your own questions for hours on end.

Instead of accepting his introspective tendency, Mark berated himself on the page for his hesitancy to engage the moment. A college essay in which he tried to disprove the famous Cartesian dictum, "I think, therefore I am," suddenly devolved into grammarless exclamations that must have been directed more at himself than to any reader: "Thoughts that exert no influence influence influence influence are not real."[24] Mark's inclination to feel himself alienated from others had followed him from Stony Brook to this tropical bohemian campus, and it lent emotional urgency to his philosophical excursions. In his dorm room, he wrote letters addressed to himself (a habit his father practiced) about being "afraid of people in the daylight."[25] He fretted over what might happen if he got too close to his peers, fearing "that they should cause my heart to soar so high, touching the sun, for an instant, and then plummet to the earth, digging a hole in the process . . . my mind's hole."[26] He willed himself back from the verge by ending these letters with some kind of reminder, à la Heidegger, that his mind was an extension of the world, not a void.

Heidegger's criticism of Descartes caught Mark's attention. When Descartes located the essence of human life in contemplation, he took it as a given that the stream of his consciousness was nearer to reality than the fleeting impressions of physical objects that passed through his senses. Descartes's learned skepticism about external objects echoed Plato's "theory of forms," which had long ago infused the material world with an air of unreliability. By elevating the thinking individual over and apart from earthly matter, Western philosophy enshrined a way of being in the world that drove an illusionary wedge between the inner life of human subjectivity and everything else.

By contrast, Heidegger argued that people and things were fundamentally "entangled." It was this idea that Mark loved most. Heidegger had flipped the script on Descartes: "I think, therefore I am" was discarded in favor of "I am, therefore I think." Or, more precisely, "I am in the world, therefore I think." The nature of our existence, according to Heidegger, was best understood as a thoroughgoing dialogue between humanity and materiality. Descartes and his ilk reduced the diversity of existence when they asserted that contemplation was the essential posture of being. While Heidegger was no stranger to contemplation and its merits, he believed that action—engaging one's body and mind to perform tasks—preceded meditative thinking. In the midst of action, we see the world much differently than we do in our contemplative mind's eye. Heidegger wanted to expand the realm of philosophical inquiry so that it spoke to the everyday, task-oriented lives of farmers, artisans, and others who worked on the land or labored with their hands.

Mark amassed a small ream of notes on *Being and Time*, which he composed in a cramped cursive that must've left his hand aching during the walk back to his dorm. He did not command a preternatural understanding of Heidegger's work in all its density; his professors' comments in the margins of his term papers dispelled that hope. But he grasped the larger argument brewing in Heidegger's talk of tools. To think seriously about the simple act of hammering a nail, for instance, was a way to put the contemplative mind back in touch with its earthly matrix. Carpenters had a very different relationship to the hammer than a gazing philosopher would. The proper way to converse with a hammer, or to divine its essence, was by *using it*: "The less we just stare at the hammer-thing, and the more we seize hold of it and use it, the more primordial does our relationship to it become," wrote Heidegger.[27] No amount of speculation could approximate the perspective you gain by holding a hammer in your hand. Carpenters regarded each material they worked with in relational terms: hammer, nail, and wood were implicitly linked when you approached them with the intention to build. To craftspeople, the sight of raw materials did not lead the mind to contemplate the Platonic ideal plank of wood—their impulse was not to ponder the abstract essence of a lone object. The things of the world were fundamentally entangled with other things and the tasks you could achieve with them using the tools you had. And all of it, moreover, was entangled with the lives of others. Grasping any artifact connected you to the mental and physical efforts of the people who built it. Tools, like Heidegger's hammers, were what forged this web of object-based connections by giving people the means to impart lasting marks on the materials at hand. Within Heidegger's worldview, the Cartesian notion that we are each "an autonomous thinking thing" appeared tenable only under the most self-aggrandizing kind of tunnel vision. Isolation was an illusion, though one hard to break free of.

A pair of hand-drawn images, which Mark later sketched during his tenure at PARC, would gesture back to the decisive impact of his youthful study of Heidegger. These two images each served to visualize a distinct theory about how humans inhabited the world. The first image—which Mark labeled "Wrong"—portrayed the Cartesian mind: the word *person* was enclosed in a perfect circle, and around the

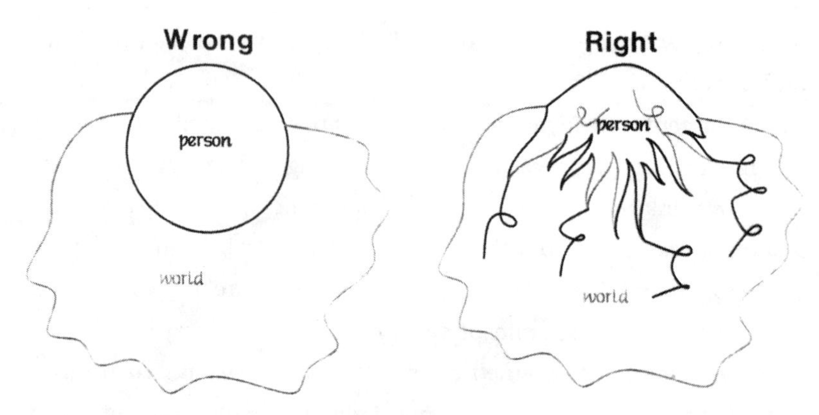

Two presentation slides, drawn by Mark Weiser in 1989, each meant to illustrate a model of how people inhabited the world. Courtesy of the Department of Special Collections, Stanford University Libraries, Stanford, California.

circle was a shapeless mass, which Mark had designated "world." A person moved through the world, within this model, as an impenetrable and self-possessed entity. In the second image—labeled "Right"—the amorphous blob for the world remained, but Mark deformed the neat "person" circle into something of a jellyfish with its underside spiraling out like flailing tentacles, fully entangled with the world. Whereas the perfect circle stood in for Descartes's theory, the jellyfish drawing expressed the ontological fluidity underlying Heidegger's philosophy. Person and world were more of a piece in this image.

Two decades after New College, when Mark was on the cusp of technical fame in the 1990s and delivering keynote presentations, he would deploy these two diagrams as a foundation for explaining the philosophical roots of his vision for computers. He would indicate how desktop computers were Cartesian—the machine effectively walled off users from other people in their surroundings—and then he'd unveil his Heideggerian blueprints for new kinds of computers that might function in step with the social rhythms of everyday life. The back-of-a-napkin aesthetic of his projector slides belied their deeply personal history. Mark had struggled through many days at New College with an inkling of these opposing diagrams that, in retrospect, appeared so emblematic of the existential revolution he waged on himself during college. From the day he discovered Heidegger in the campus library, Mark would strain constantly, from adolescence through adulthood, to force his Cartesian eyes to see the world through Heidegger's lens. The technologies he would go on to invent were, in some ways, an attempt to continue the personal transformation that he had undertaken during college.

But despite Mark's first blush of enthusiasm for Heidegger's ideas at New College, entanglement proved much easier to understand than to experience. The joyous epiphanies of his reading did not translate readily to his normal life, especially when it came to people. He still felt distant from his peers on campus, even after the nightly comradery of playing go by the pool. From his dorm, Mark dreamed up scenarios in which strangers could read each other's minds. Fantasies of extreme interpersonal connection pervaded several pieces of fiction and poetry he wrote at New College.

In one short story he titled "Hello," Mark's protagonist, John, meets a stranger who grants him the power to see thought bubbles floating above every pedestrian he encounters. Mark revised the story just two days later, doing away with the unnamed stranger. In this draft John gains the ability to read others' minds simply by exchanging hellos with them, whereupon they can read his mind as well. John soon realizes that "when he had read someone's mind he found it impossible to dislike them," and proceeds to befriend every oncoming pedestrian whose mind he reads.[28] The story's quaint moral hinted at the stresses of an eighteen-year-old who frets over the question of how to greet people he encounters: *Nod or smile? Now, or wait until we're closer?* If only we could all make ourselves transparent to one another, the story laments. If only the pulse of our inner lives could beat up to the surface. In all the fiction Mark wrote that year, an odd punctuation usage appeared to reflect the psychic pull of this wish. He regularly used quotation marks whenever he recorded a character's stream of consciousness, presenting their unspoken thoughts in exactly the same format as their spoken dialogue. This device seems to have been less a deliberate style choice than a subconscious tic attesting to Mark's longing for fluid interactions and lucid relationships.

Mark's most listless collegiate poems and stories were written on Mondays. They sagged with a subtext of weekends amounting to nothing. In one Monday poem, he wrote, "I want to feel someone under my arm who wants to be there, and be under someone's arm who wants me. Is there any reason these things should be so hard? Why is it that I cannot just put a sign up 'Love Wanted' and be tried out?"[29] Yet Mark projected mixed signals about his stance on companionship. He made a point of withdrawing to think and create, he skipped more classes than he attended, and more than once missed several weeks of school while hitchhiking back home to Long Island. The protagonist in one of his untitled Monday stories walks into a deli and almost has a nervous breakdown as he subjects himself to the apparent anguish of waiting in line for a ham sandwich. The character then trudges back home to sit and wonder why innocuous gatherings of all sorts cause him such anxiety. Mark, lacking any answers to offer his character, reached for an epiphany in the form of a harried soliloquy, formatted as dialogue: "Life with troubles every time I go outside my door is just not worth the trouble. But I have to go outside my door sometimes if I want to keep living. I don't know. Fuck, maybe I'll try partying or something tonight."[30]

Painful as they were, Mark's literary grievances functioned as pep talks. In them he rehearsed the transformations he might undergo. They cleared paths for his potential development from a fretful boy into a man at home in any situation. Progressing from Descartes's incessant introspection toward the worldly entanglements of Heidegger's *Being and Time*, he might grow to awake with the conviction that each new morning and every passing moment belonged to a grand expedition taken by all humanity. As he glanced up at the rising sun, he would feel present to everyone who was also gazing upon it, and feel their presence. He would envision another person's hand on the door handle when he opened it, as if that hand were grasping the handle in unison with his own. If he couldn't turn his peers into Heideggerians, then at least he could try projecting imaginary communions onto the objects and places they all beheld. This budding sense of togetherness might spring into relief like his shadow in the night, appearing suddenly in the illumination of a lamppost amid the desolate quad, only to remain hidden in the routine light of day.

Mark left New College after just three semesters. His mounting tuition debt scared him more than he had anticipated. He missed Reich. The delight he sought in professors' comments on his work, a passing respite, no longer seemed worth its cost. On the occasions he attended classes, Mark gracefully chimed in, addressing his instructor more like a colleague. In academic settings, Mark was so obviously the son of a professor, and the familiarity of scholarly rites had rendered them banal. Only the allure of working alone on his own whims kept Mark engaged, and his transcript bore six "incomplete" courses to show for it. In a letter attempting to explain himself to the college's provost, the young Weiser brushed aside an abandoned course on the strand of mathematics called real analysis, saying, "The class gave me some understanding of where Analysis was going, and so also enabled me to decide not to follow that path at that time."[31] A few professors noted embers of brilliance when evaluating Mark's intellectual promise, though none appeared to respond to the letters he wrote them after he had quietly dropped out.

But if Mark felt like a failure as he cleared his dorm room, he didn't pack like one. Everything he wrote, every poem and every paper, was neatly stored away for future reading. He would move often over the next decade, and this hodgepodge of youthful prose always made the trip. In December 1971, Mark took to the highway with all his belongings. Perhaps Ann Arbor would hold an answer. He had old friends there. The Midwest was a bit closer to home, and to Vicky Reich. A hesitant flick of the turn signal began the northbound slog into another season, as he sped past breeze-swept greenery for the dead of Michigan winter.

To deem Mark's work as a philosophy major a success would be a stretch. His yearlong struggle to wrestle himself from a Cartesian worldview to a Heideggerian one was, however, a very formative quest. He knew Heidegger was right. In any case, he knew he didn't want to continue feeling like it was natural to be insulated from the world. He had tried hard to refashion his relationship to the people and places around him. All these efforts played out in secret, though, surfacing only in words shown to no one. Each new thought experiment carried the burden of being crafted from scratch. Last night's insightful sensation left no trace on today's happenings. The tonic of real connection would not become a stable part of his life until he managed to create something he could share—something that made a mark outside of himself, some kind of tool that helped people attune to all the manifold, tacit conversations they might have with their surroundings. "The infinite richness of the universe," as he would phrase it much later, was always there wherever you were, teeming beneath the hazy chatter of self-talk.[32] But Mark had come to another realization at New College: he couldn't bridge the gap between theory and practice on his own.

Mark fell back on computers in Ann Arbor. Programming was, after all, an efficient way to make money. Soon after leaving New College, he took a part-time job earning full-time wages at a small firm called Omni-Tech, where he wrote code for industrial computing systems that helped Detroit-area companies manage their immense factories. The job allowed him to pay off his tuition debt and, most important, left him with many spare hours to read, as well as to try branching out into an alternative community he had read about.

In the fall of 1974, three years after quitting college, Mark composed an "audio letter" to his father David, which incidentally gave Mark pause to reflect on the man he was becoming at age twenty-two. Alone in his apartment on a windy night in September, Mark spoke into a microphone for an hour, recording the equivalent of an hourlong voicemail on a cassette tape that he would mail back to Stony Brook. Speaking to his father in this monologue format was perhaps not so different from talking to him at home. Mark had grown comfortable with the silence and learned to press on in the belief that David was listening in his own way. Mark assumed the burden to be interesting, the way a lecturer on stage talked to a seated crowd. Throughout the recording, he carefully distilled the takeaway points associated with his anecdotes and worked into the flow of his remarks the questions that David might've thought to ask.

Audible in between these scholastic pleasantries were sounds of the joy and good humor that now suffused Mark's life in Michigan. He intimated to David how his love for Vicky Reich was deepening in spite of the distance that still lay between them. Reich was completing her biology degree at Gaucher College, near Baltimore. They had continued to visit each other over the summers and holidays during her undergraduate years. Mark noted in the recording that he had begun writing Reich postcards every day. Soon she would move to live with him in Ann Arbor, and to start a master's in library science. Speaking of his own academic status, Mark giggled as he told his father about the computer science department at the University of Michigan, which had just admitted Mark into their graduate program even though he lacked a bachelor's degree. Mark highlighted the differences between his new coursework and his programming work at Omni-Tech, and said he planned to quit the job soon. "I'm just not growing there anymore," he confessed into the microphone. "You have to sort of be a loner . . . going off and creating a computer system."[33] His present interests lay in creating ideas about computers.

He spoke of his growing enthrallment with the pleasures of basic research and the intellectual communions it beheld—"the contact with people, thinkers and so forth, or 'great men,' as you've often said to me, Dad."[34] He described the gist of a technical paper he had just written, which put forth a design for "live-time computing" that would give users a running display of the tasks their computer was processing and how close each task was to completion.[35] Making this information readily visible, Mark reasoned, would avail people from having to sit staring at the machine unawares, wondering if it was executing their commands or not. He didn't want users to wait around without a sense of what was happening under the hood, the way he used to wait around for his punch cards at Stony Brook's Computing Center, a wall standing between him and the mainframe. The paper promised to be a minor contribution, he acknowledged, but it was the first step of an alluring line of thought.

Since leaving New College, Mark had begun to wonder how computers might become more usefully entangled with people. He wanted to create applications that showed people exactly what they might want to know at a given moment, in the knowing manner of a friend. It was all very nebulous still, but already Mark sensed that he would grow into a different kind of technologist than the sure-minded mathematicians who surrounded him in the campus computer labs. "I really believe," he uttered into his little microphone, "that the philosophical side of my ideas is more earth-shaking, as it were, and that is really where the difficulty is."[36] But remarking on his academic trajectory was not the main impetus behind Mark's audio letter to his father. Apart from his classes, he was pursuing an emotional education off campus. This extracurricular training gave him the nerve to end his recording on an unexpected note.

Suddenly, Mark's voice softened and slowed into the rhythm of a trance. All his remarks leading up to this moment now felt like a perfunctory opening act. They were merely echoes of the diffident son he had been before. Mark had recently learned, after a year of evenings spent intimately with Ann Arbor strangers-turned-confidants, to dwell deep within himself and to express, almost casually, whatever feelings he discovered there. He knew his father would likely be listening to his cassette in the Stony Brook hospital bed where he was being treated, lying weakened from a decade of alcoholism. "It's very exciting for me," Mark said in his altered, mystical tone, "to see you so . . . fine. So very full of life, and handling your life and in control of your life."[37] He maintained the gentle, run-on cadence of a yoga teacher. "To have you, Dad, as a model, as someone I can look to and say, 'Yeah, I wanna be like him—look at this guy, he's done really well, and he knows himself and he knows the world, and if I could be like him that would be . . . enough.'"[38] After a pause, Mark countered, "And I'm not like you." But, pausing again, he abandoned that notion and continued to regale their affinities. "We understand that there is to the world a mystery, an ineffableness—something which hides itself, to make itself known. And this is not intolerable to us, as it is to some people. . . . Instead, we know how to let go to this, to let it let ourselves go and have it move our thinking. . . . It's letting go into the mystery that is the most human activity."[39] Mark seemed to be thanking him, forgiving him, and maybe on some level telling him goodbye.

Why Mark's father would be the recipient of this tribute makes sense only in light of the somewhat experimental, psychotherapy-adjacent group Mark had joined recently in Ann Arbor. His newfound manner of expression had been fostered at gatherings with a random assortment of like-minded devotees. They met in community centers or church basements on weekday evenings. Some came straight from work. A saleswoman might take her seat next to a mailman, who was seated next to a college student and middle school teacher. Huddled around in a circle, Mark seated among them, they would then pair up for each session. They attended every meeting on the promise that someone would be there for them—maybe someone new, but the protocol would remain the same. The encounter began with a few minutes of small talk before they each submitted to the practice.

Evenings at the Ann Arbor chapter of reevaluation counseling, like all the other chapters forming across America and Europe, were choreographed from afar by a man in Seattle. Harvey Jackins founded the organization during the 1950s, after a life-changing month he spent consoling a distressed coworker.[40] The guy was awash in his own despair, and Jackins's attempts to engage him in rational dialogue failed, so Jackins resolved to just listen closely while his coworker ranted and cried. It turned out that this was exactly what the guy needed. Once he had divulged to Jackins the fury of his inner turmoil, he no longer felt suffocated by it. A theory started to flutter around in Jackins's mind. Perhaps it was misguided, or even counterproductive, to insist on workaday stoicism in the face of class conflict. A lifelong labor organizer, Jackins had been a Communist activist at the University of Washington where he studied Marxism and Maoism. His efforts to fight against worker exploitation had landed him a subpoena to testify before the House Un-American Activities Committee.[41] As Vietnam War–era activists were taking to the streets, Jackins began to invite more of his comrades into his home. The battle to end oppression, he believed, must start with a communal initiative to confront the experiences of oppression that the world's have-nots had tragically internalized. Seeking to replicate the initial success he had with his coworker, Jackins used his home as a lab for testing out new counseling techniques and he soon opened a small office in downtown Seattle. The movement spread beyond the city by the late 1960s, as Jackins fueled its expansion by self-publishing a small library of manuals and books to introduce new members and workshop leaders to the homespun fundamentals of reevaluation counseling.[42]

This was the version of the movement's origin story that enticed Mark. By 1975, he had become a fixture in Ann Arbor's reevaluation counseling community and initiated a penpal relationship with Jackins that would last for nearly a decade. In his correspondences and speeches, Jackins downplayed his prior involvements with Scientology and the ways in which his counseling techniques were clearly informed by the pseudo-philosophy of emotional release developed by Scientology founder L. Ron Hubbard.[43] Whereas Scientology subjected its members to "auditing sessions"—wherein a church-appointed auditor compels participants to take inventory of past distresses that still weight on them—Jackins sought to uncouple this tactic from Hubbard's grandiose religious project and couple it instead with left-wing liberation movements then gaining traction. He reached out from his roots in labor unionizing to include gender and race. While liberal white people constituted a majority of the organization's growing membership, reevaluation counseling was billed to be a nurturing haven where people came together to heal one another from the chronic miseries wrought by capitalism, sexism, and racism. Mark bought in quickly to Jackins's premise that intimate conversations held the keys to personal growth and social change. Several pages in Mark's Ann Arbor notebooks bore the heading "What Harvey Says." He drove hours to hear Jackins lecture whenever the man came to the Great Lakes region. Mark recorded the nuggets and promises that Jackins gifted to his committed listeners: "There is nothing more revolutionary than people interrupting oppression one-to-one as it happens. . . . This will guarantee the collapse of the oppressive society."[44] And yet, if the tyrannies of oppression were what compelled him to join Jackins's movement, it wasn't exactly clear which crosshairs Mark felt himself to be in the middle of. He had read and admired *One-Dimensional Man*, Herbert Marcuse's famous 1964 critique of capitalist bureaucracies, and he would become a vocal ally for women in Ann Arbor and later in Silicon Valley. But Mark's interest in social justice was less pronounced than his acute, aching need to feel really present with other people.

More salient than Jackins's politics, it was the routine practices of reevaluation counseling sessions that spurred Mark's involvement. The movement was spreading rapidly—particularly in college towns—as a grassroots alternative to psychotherapy where all were welcomed. Instead of lying on a licensed psychologist's couch, "RCers" turned squarely to each other. Whoever Mark chose to partner with on a given night would listen closely as Mark labored to relive painful experiences from his past. If Mark showed any sign of withdrawing into himself as he narrated the memories, then his partner would ask a pointed question or make an encouraging gesture to recalibrate Mark's effort to share his anger and his sadness. It was the partner's job to facilitate Mark's "emotional discharge," which would (ideally, if seldomly) be punctuated by bouts of crying, yelling, trembling, and hugging.[45] On at least one occasion during the fall of 1975, the subject of Mark's discharge had been his father. In a scribbled notebook entry about the session, Mark summarized the focal points of his partner's remarks as well as his own: "Mark: afraid of his father hurting, then seeing him helpless brings tears."[46] Whatever emotions happened to surface, the process asked participants to take turns giving each other their undivided attention. The ensuing sense of copresence often proved quite restorative, particularly during a decade when Americans were spending more time at work and devoting many hours to their televisions. Reciprocal attention was a profound gift, and Mark would be wary of emerging technologies throughout the 1990s that required people to stare at screens.

Jackins had his own theory to account for the practice's effectiveness. As people matured through childhood into adolescence, he explained, they were pressured by parents and teachers to temper any impulse to whine or lash out at the onset of displeasure. Thus, as a result of learning to downplay pain, people habitually internalized the bulk of their negative feelings and, from here, hatched patterns of thought conducive to self-loathing, defensiveness, and an overall reluctance to open up to others. In the words of a standard handout Jackins had written for all first-time participants: "[Reevaluation counseling] theory assumes that everyone is born with tremendous intellectual potential, natural zest and lovingness, but these qualities have become blocked and obscured . . . as the result of accumulated distress experiences."[47] Mark and his fellow RCers came to each meeting prepared to share their feelings with the conviction that unbridled release would free them to think more clearly and joyfully once they had ascended from the church basement and walked back into their lives.

While the nightly sessions revolved around the back-and-forth of emotional discharge among attentive partners, these venting exercises also served as a means to a broader end: the ultimate goal was to render yourself intelligent, in Jackins's peculiar sense of the word. You yelled about the time your father lost your favorite toy, or cried about the time your boss reprimanded you for no valid reason, because doing so would—according to Jackins—allow you to be more rational as you went about your day. Jackins's theory of intelligence, whether he knew it or not, resonated with the anti-Cartesian line of philosophy that still dominated Mark's bookshelves. There was something vaguely Heideggerian about Jackins's assertion that "the essence of rational human behavior consists of responding to each instant of living with a new response, created afresh at that moment to precisely fit and handle the situation of that moment."[48] Jackins's crowning self-publication, *The Human Side of Human Beings*, was to Heidegger's *Being and Time* what a can of spam is to a slow-roasted ham. Jackins served up the essence of humanity by first explaining how we differ from "non-living matter" and then, with a few more paragraphs, how we differ from nonhuman creatures.[49]

The most interesting aspect of this three-pronged theory of everything was the basis he established for making his oversimplified distinctions. Jackins argued that the fundamental difference separating entities in his three categories was the respective way they inhabited their environment. Referencing billiard balls and chairs, he observed that nonliving matter is essentially passive in its relationship to all other matter. "Give a chair a push and it is pushed," he wrote.[50] Inert objects defer to the will of *the living*, who go about altering the environment to better suit their needs. Nonhuman creatures, Jackins posited, are bound by their instincts when they interact with one another and with nonliving matter. An animal's instincts—"pre-set patterns of response . . . fixed in the heredity of the individual creature"—make for a very limited playbook of possible actions that each animal must select from as it responds to the various situations it encounters.[51] Some animal and plant species have evolved to acquire more patterns of response than others, but none can intuit an entirely new behavior. A codfish, to cite one of Jackins's examples, "is able to respond to the environment only in codfish ways, not in ivy vine ways or butterfly ways."[52] Only we humans could transcend the default settings of birth: "The human being can and does continuously *create new* responses all through the lifetime of the individual."[53] While the capacity to act on instinct is instrumental to the survival of nonhuman creatures, Jackins regarded knee-jerk reflexes among humans as an indicator of grave deficiencies. Human beings who proceeded through life with "rigid, pre-set responses" were, in Jackins's schema, forfeiting our great defining quality. Being "flexible" and "creative" in our ongoing interactions with various environments was the essence of humanity and the engine of progress.[54] He ultimately derived a theory of intelligence from these grand comparisons between humans and other creatures. Intelligence, according to Jackins and the RCers who studied his teachings, was synonymous with a person's capacity to "create an endless supply of new, tailored-to-fit responses to the endless series of new situations we meet."[55]

Mark strove to nurture this capacity as he led his own reevaluation counseling workshops in Ann Arbor, and Jackins's emphasis on contextual prowess would retain a high place in Mark's mind when he eventually set about designing responsive technologies at Xerox PARC. This picture of intelligence ran counter to more familiar images of braininess in science and technology, as well as in the arts and pop culture, like the brooding philosopher depicted in Rodin's sculpture *The Thinker*, or the stock mathematicians who appeared in Hollywood films, standing in their chalk-dusted sweaters before a wall full of equations. These figures personified a notion of intelligence that was presumed to reside in deep recesses and remote corners, set apart from society's surface-level hubbub. The dictums that Mark subscribed to, by contrast, maintained that intelligence was a *relational* phenomenon. Insights came when humans brought their minds to bear, ever so closely, on the unique realities unfolding around them in the present moment. Your surrounding environment levied a decisive impact on your state of mind, according to the reevaluation counseling teacher training materials Mark studied. The teachers were instructed to regard the design of their workshop spaces as an essential component of the therapeutic mission: "Anything you can contribute or do to make our surroundings more interesting or beautiful is most welcome."[56] A stimulating place might keep the mind from turning in on itself.

Jackins liked to compare the mind to a computer full of files documenting past experiences. Information gathered by your senses was processed and compared against an interior database of retained impressions that called up memories most relevant to the immediate situation. Every new experience—sometimes referred to as "information" in Jackins's verbiage—was inevitably framed, shaded, and colored in by this network of personal *data*. "The incoming information," Jackins wrote, "is *understood* in relation to other information, in its similarities and differences to other data, not ever as a concept by itself."[57] Data associated with past experiences constituted a basis for making sense of and responding to present information; however, no two experiences were ever perfectly identical. True intelligence, RCers learned, was a balancing act. You wanted to amass a useful database of past experience so as to have as many reference points as possible at the ready. However, each reference point—each bit of remembered data—was valuable only when it augmented your awareness of what was happening in the here and now. In a note Mark scribbled about this subject during a workshop, he wrote, "Boredom is too little information or too much so that we can't connect it to anything."[58] Under Jackins's model, not only could your intelligence and well-being be limited by a dearth of retained knowledge, it also could become constrained by your habit of mindlessly revisiting memory files that bore little relation to the current situation. Your intelligence, along with your quality of life, diminished whenever your attachments to stored information distracted you from achieving the serene elasticity of a discharged mind.

Never more than a metaphor in Jackins's theory, the computer still occupied Mark's working hours as both the subject of his research and a hulking contraption he operated daily. On top of attending his graduate classes and leading reevaluation counseling workshops, Mark had happily taken on two additional roles. In December 1976, he and Reich were married, and in June 1977, Reich gave birth to their first daughter, Nicole Reich-Weiser. Midway through his computer science PhD degree, now with a newborn to help support, Mark got a job counseling undergraduates who were trying to write their first program in the University of Michigan Computing Center, home to the illustrious Michigan Terminal System (MTS).

Launched in 1967, the MTS had become one of the world's preeminent terminal-oriented time-sharing systems by the mid-1970s. Those with access to this system no longer needed to create programs on punch cards and hand them in to the mainframe operator for batch processing. MTS users sitting at a workstation of their own issued commands directly to the center's mainframe computer and saw the results almost instantly on their display. At first, this form of "interactive computing" took place on teletype machines, which made a running printout of each user's dialogue with the mainframe and supported up to forty users at once.[59] The system quickly grew into an intercampus network that spanned eight universities across the United States, Canada, and England. Students and faculty at each campus could share files with those at other campuses, as well as post messages on discussion threads using CONFER, a program developed for the MTS in 1975 by UM graduate student Bob Parnes, and which would influence more-famous internet forums like the Bay Area–based Whole Earth 'Lectronic Link platform. This connectivity facilitated remote research collaborations and even an online community of sorts: "Users gave one another tips and advice, pieces of code and, at times, pieces of their minds."[60]

As the MTS user base grew to the thousands, the University of Michigan Computing Center also acquired new terminals, such as the Tektronix 4010, which electronically displayed neon-green texts and graphics in motion—a stunning advancement when viewed alongside the decade's teletype machines and text-only terminal screens.[61] Compared to Mark's earlier computing experiences at Stony Brook, Michigan's center housed a digital wonderland of cutting-edge hardware and software. Campus programmers interfaced directly with the mainframe computer, allowing them to write, test, and debug their code all during the same session. Users of all levels could log on to several virtual spaces and converse with counterparts who lived far away.

Amid the patter of keystrokes and electronic noise, Mark maintained a watchful eye over the students, each sitting at their own workstation, as they endeavored to create a working program. His own programming work had always kept him entrenched in his code and the tasks he wanted it to execute. As a computer counselor, he now attended to other people, looking on as they wrote code and noticing how they engaged with their machines. The act of computing, so dynamic and complex when your fingers were on the keyboard, took on

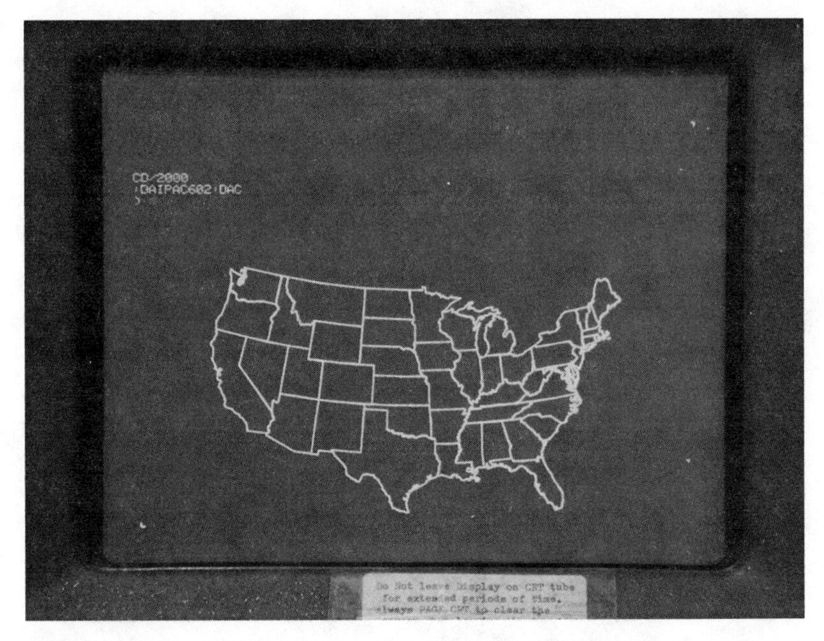

A Tektronix 4010 terminal displaying graphics. Photograph by David Gesswein.

a rather mundane quality when you sat observing rows of other people typing, pausing, squinting at their screen, and typing again. A room full of people interacting with their respective terminals made for a vastly different scene than the one Mark occupied through his other counseling role in Ann Arbor. Students came and left without paying the slightest attention to their peers. Michigan's time-sharing system was captivating—its speedy responsiveness, its repertoire of shared files and online knowledge, its networks playing host to the electronic activities of so many people at once. Mark would often, like the students he helped around the lab, pass hours with his eyes locked on the screen when he wasn't on the job. He knew its charms and its powers, but he also knew—from reevaluation counseling—the feeling of giving the person next to him his complete focus and having that person devote all their energy to him as he spoke. The intense copresence he aspired to experience in the evenings among his fellow RCers must have made the stark absence of mutual regard among Computing Center users all the more apparent.

Students at work at the University of Michigan Computing Center. Item no. HS10537, courtesy of the Information Technology Division Records, © Bentley Historical Library, University of Michigan, Ann Arbor, Michigan.

While time-sharing terminals linked users to new bounties of networked information, those users had to disengage from all sorts of other environments in order to concentrate on their machines. Greater connectivity begat greater immersion; immersion, viewed from another angle, begat isolation. The circulation of data stored digitally remained limited to heavily wired, windowless rooms. Elsewhere—at reevaluation counseling workshops he led, in his graduate seminars, around the house with Reich and their little daughter Nicole, on walks through the woods—Mark sought to commune as deeply as he could with the people and places he encountered. He had moved to Ann Arbor adrift on a whim, and his life since then had quickly changed. There was so much for him to marvel at, so much he might be missing.

Jackins's relational definition of intelligence, Heidegger's sense of entanglement, and Polanyi's conception of tacit knowledge each galvanized Mark's intellectual development, but neither card punch machines nor time-sharing systems seemed amenable to such notions. How intelligent was it, in Jackins's terms—or Polanyi's or Heidegger's, for that matter—to devote so much of your body, mind, and being to the electronic information confined on a desktop monitor? When Mark looked around the Michigan Computing Center at all the students looking at their screens, no one else appeared to be concerned.

By the spring of 1979, soon to be a newly minted PhD with job offers in hand from the University of Maryland and Duke, Mark had set himself up for a promising career as a computer science professor. He chose Maryland, and the family settled in Silver Springs, where their second child, Corinne, would be born. At Maryland, Mark embarked on a life of days spent among computers with students in a lab and working solo at the desktop in his office. Meanwhile, the rise of smaller, more personalized computers—like the Xerox Alto and the more commercial models it inspired—would draw cadres of new converts to immerse themselves in the pixels of interactive screens.

Asymmetrical Encounters

ZOTT'S BEER GARDEN COULD SEEM LIKE an odd place for computer scientists to sit and ruminate about the future, but it wasn't. Its picnic tables sat on a lawn of woodchips nestled under a thick roof of trees in Portola Valley, just west of Palo Alto. Kids and dogs roamed at will, and every hour or so, a new customer might arrive on horseback. Anyone who has lived among the sprawl of tech-company headquarters will have noticed tech workers' occupational fondness for rustic charm. When they leave their labs, they head to the woods, a backyard woodshed, or a place like Zott's. A small plaque near the entrance informed Zott's patrons that it was the site of an experiment that launched the "beginning of the Internet Age." In August 1976, a team of scientists from the Stanford Research Institute got permission to set up a computer terminal on one of the shaded picnic tables. Cable wires connected the terminal to a van full of electronics in the parking lot out front. From their picnic table, they sent a packet of data that traveled intact all the way to Boston, as it first passed through the antennas of a local radio network to their offices down the road and then across the country via the ARPANET. This was the first electronic message to hop seamlessly between networks, and its transmission suggested that all manner of wireless and wired networks could be fused into an overarching global system. As the Stanford scientists clinked pint glasses to celebrate their achievement, they couldn't have known how monumental that summer afternoon at Zott's would prove to be.

On a May evening in 1988, Mark Weiser's beer was getting warm and his burger was getting cold. He spoke rapidly and softly as the others drank. Gathered around his picnic table were five senior researchers who he worked with inside Xerox PARC's Computer Science Lab. Weiser had been there for almost a year now, and he was finally acting on the urge that drove him to Palo Alto. The group had shown up at Zott's on Weiser's invitation, and he had billed the meeting as an occasion to imagine what technology should be like twenty years ahead, circa 2008.

The computers presently leading the market in 1988 had first taken shape on a back-of-a-placemat drawing at a Houston diner six years earlier. From that initial sketch, the three cofounders of Compaq had pioneered the notion of a portable (or "luggable") twenty-eight-pound PC that businesspeople could travel with. Compaq chipped away at IBM's virtual monopoly as the two companies took turns introducing upgraded models designed to steal the other's customers. Compaq seemed to have recently taken the lead by inching ahead of IBM on speed, cost, and software compatibility. These marginal advantages were translating into major business victories for Compaq, but the whole battle sucked the wonder out of desktop workstations. Much of what a PC could be suddenly felt already determined. Industry winners had set the design parameters; adding innovation around the edges was probably lucrative, but also boring. That's why Weiser had called his colleagues to Zott's.

Their meeting generated no prophetic doodles, and it failed to merit a commemorative plaque in the beer garden. Only one thing came of it: everyone agreed that the conversation was worth continuing, the timing was right, and Weiser should lead the effort. At the others' insistence, Weiser said he would include everyone at CSL, which meant that a picnic table would no longer suffice.

The Computer Science Lab staffed nearly fifty technologists, all of whom believed themselves to be on the brink of eureka. This one shared attitude unified the lab more than anything else. Infringing upon the autonomy of anyone else's research saga was a sin. Being a part of this lab meant you were above mandates—you had earned the privilege to work on what you wanted when you wanted. Few arrived at their office before 10 a.m., and many stayed after 10 p.m. Most came and left as they pleased, except for the two sessions per week when the staff met as one: a "high tea" social hour, and a weekly talk by one of its members. In keeping with traditions set by the lab's founding cohort, it could still be a contentious scene; everyone was brilliant and free and ferocious. Researchers took a turn at the podium, week after week, nursing a faint hope that their pet project might illuminate some vast, essential wonderland of interlocking gadgetry around which the whole lab would rally, as their predecessors had with the Xerox Alto during the 1970s. But at minimum, each scientist just wanted to survive the Q&A session without feeling like their life's work was shoddy, inconsequential, second rate, or any other sleep-robbing adjective the crowd might whisper. Third-party appeals did nothing to help one's cause. Any mention of customers, marketing, or sales projections during the early stages of research was tantamount to pandering. Ideas were judged only by the beauty they evoked in highly guarded technical souls.

Weiser had not worked there long enough to fully grasp the lab's communal devotion to self-regard. On June 14, he assembled the staff in PARC's auditorium, hoping to pick up where his picnic-table group had left off a month earlier. Weiser opened this forum—the first in a series of "visioning meetings" he planned—with the same question he had raised at Zott's: "How do we want the world to be like in 20 years?"[1]

PARC was a place of ambitious plans and long-term thinking, of course, but even here a question like that seemed too abstract to be useful. Computer scientists were often too busy building electronic worlds to tinker around with "the world." Plus, twenty years may well have been a hundred within the industry's exponential time warp.

Weiser pressed on. He explained what it meant to have a vision and why the Computer Science Lab needed one. Visions, he told his peers, "are descriptions of a desired state. . . . They are directional, elusive, idealistic."[2] The absence of a vision bothered Weiser. From his decade leading reevaluation counseling workshops, he retained the ritual of writing down his goals weekly on a detailed chart. Every week for years, Weiser and the other RCers had jotted down personal goals for the coming week and then evaluated their ongoing goals already penciled in for the next month, the next year, the next five years, and the next twenty. In addition to this stack of personal goals, the chart included four additional columns for goals associated with "My Family," "My Allies," "For Mankind" (which Weiser amended to "*Hu*Mankind"), and "For All Living Things." The chart thus tasked its users with twenty-five goals in total. On a chart he had completed back in Ann Arbor, Weiser had written a distant goal that wasn't far off from those he now wanted to pursue at PARC. His wish "For *Hu*Mankind," twelve years before, had been for everybody to prioritize "friends over machines."[3]

After Weiser concluded his opening remarks, he split the lab's staff into breakout groups and handed each a discussion prompt. It began with the following premise: "Our current computer environments do not open themselves to people, but they could."[4] Using a desktop was for most people like trying not to drown in "a sea of bits," Weiser said.[5] The machine demanded an untenable burden of total immersion. "These days," he continued, extending his analogy, "we enter our computer systems briefly, holding our breath, never totally comfortable, and always glad to come up for air [back to] real-life."[6] Weiser declared that the "integration of people and the sea of bits" was the seed of a budding vision that he'd be sharing with his breakout group.[7] It was a telling moment, in retrospect—indeed, some CSL scientists later suspected that these meetings were just Weiser's way of politely pressuring them to adopt goals he already had in mind.

Each group reported back with some enthusiasm but nothing in the ballpark of a unifying vision. The first offered their speculations regarding the fate of literature in an advanced electronic age, as if no technological vision could be complete without tackling that issue. The second group tried to summarize a discussion they'd had about game theory and geometric algorithms, and about parallel algorithms and approximation algorithms. Another group wrote and performed a one-act play called "The CopierLess Office," in which a reporter from the *San Jose Mercury News* interviewed a photocopier.[8] The final group compared the merits and flaws of several programming languages.[9]

Weiser had worried the visioning meetings would go like this. He had asked the groups to weave a tapestry from their various desires, and they returned instead with a cubist collage of predictions, jargon, bad theater, and old squabbles. Before the next session, he sent out a lab-wide memo reminding groups to synthesize their ideas rather than just "stapling them together."[10] The memo had no discernable effect on subsequent meetings. His colleagues talked past the problems of personal computing that Weiser intended to rally their attention around with his aphorisms and metaphors. Everyone had a talent for seeing a separate future that was unmistakably related to the research each was currently steeped in. Perhaps the only redeeming thread of discussion, tangled up in the digressions, was the passing references to a few projects the staff had collaborated on recently.

One of those projects stood out to Weiser. A year earlier, Bob Sprague and Richard Bruce had circulated a proposal to build wall-size screens, and the thought of these "computer walls" still excited researchers across PARC's various labs. Sprague and Bruce figured that, given enough time and money, they could make screens do more than display graphics. A big screen mounted on the wall might convert the room itself into a computer of sorts. Rather than typing on a keyboard connected to your desktop monitor, you might walk over to a computerized wall and handwrite your words directly on the screen with a special electronic pen. By virtue of its size and placement within an office, the computer wall might prove more communal than the personal computers in each employee's workspace. There was something Xeroxy about that. Like a photocopier, it seemed like a machine people would gather around and share. Sprague and Bruce believed they could also endow the screen with the capacity to scan documents. Office workers on their way to the water cooler would be able to make an electronic copy just by slapping a piece of paper against the wall.[11]

Sprague and Bruce's proposal resembled aspects of experiments tried down the hall at PARC as well as at MIT. To the team who had just designed PARC's "Colab" in 1986, the computer-wall initiative hardly sounded new. Already up and running, Colab was "an experimental meeting room" capable of hosting five participants.[12] Its creators were motivated by a statistic: office workers spent 30 to 70 percent of their workday in meetings.[13] While desktop computers had entered the American workplace, computing of any kind remained absent from most meeting rooms. Going to a meeting generally entailed leaving all electronic files behind, and so office workers had to print out whatever information they wanted to have on hand. Hardcopy handouts, pens and notepads, chalkboards, and boxy gray overhead projectors were the hardware of boardrooms, classrooms, and conferences.

"Media influence the course of a meeting," PARC's Colab team had asserted, in a paper calling on computer scientists to rescue white-collar workers from the banality of traditional gatherings.[14] The researchers singled out whiteboards and enumerated their shortcomings: they afforded limited space to write on, illegible handwriting rendered them useless, they needed to be erased to make room for the next use. In place of a whiteboard, the Colab team installed a jumbo touch-sensitive computer screen on the room's front wall. The screen was connected to a keyboard at a podium where the meeting's leader could stand. Facing the screen and podium was a semicircle of four desks, each with PCs that were wired to the big screen. Together, the four computers formed a "multiuser interface" enabling all the participants to make individual contributions to a shared document at the same time.[15] The whole group could contribute, read, and edit all at once. Typing usually overtook talking. The PARC scientist Gregg Foster remembered his Colab experience as "terrifically productive, to see other people's ideas popping up and being connected to what I was doing as I was doing it."[16] In Colab, meetings felt more like collaborative working sessions—more like an interpersonal extension of the work done at one's individual desk.

The Colab project reflected the basic spirit of Weiser's desire to better integrate "people and the sea of bits," as he said at his lab's first visioning meeting. But there was one snag about this experiment: the Colab team reported, after holding meetings in the room, that this system they had built to overcome the limitations of chalkboards soon felt "too limiting" itself.[17] Having desktop computers in front of them during their meetings only intensified the researchers' longing to access their personal files stored on their PCs back in their private offices. Colab's initial design sentenced each user to a session of computing without one's own electronic stuff. On the Colab workstations, no one had access to their personal email or the documents stored on the desktop in their office. The team agreed that a system upgrade was in order. Colab 2.0 allowed everyone to toggle on their screen between "public windows" (which hosted the group's live collaborative work) and "private windows" (which granted users remote access to programs on their office PCs).[18] Meeting participants were no longer limited to working together; all were now able to check their email.

Liberations of this sort, Weiser worried, might lead PARC astray from making real progress. Whenever you installed desktops in a new environment, you also brought into that space all the norms and expectations people associated with their PCs. Meetings held in the presence of PCs were, at their worst, reminiscent of sitting in a computer lab. At best, they could unify a team's creative energies like a band in a jam session; but the itch of one's personal files was just a click away, and even PARC scientists couldn't resist scratching. So while the Colab room boasted more technical sophistication than a lone computer wall, it harbored a desktop's penchant for short-circuiting shared attention. PCs seemed like the wrong device to sync up with a wall-size screen, but no other existing interfaces promised anything better. In any case, Weiser did not want his lab to do what Nicholas Negroponte was doing with computer walls at MIT.

Negroponte, a young and handsome Manhattan-born architect, had burst onto the computing scene after his dazzling talk at the first TED conference in 1984.[19] He believed that digital media would reshape the built environment and that the design of electronics could use an architect's touch. From his research post at MIT, he enticed rich investors and brilliant engineers to help him manifest his visions. Negroponte presided over his lucrative venture with the panache of a Hollywood producer and a salesman's silver tongue. Whenever he cast his spotlight on prototypes hatched in the MIT Media Lab, he dressed them up in buttery prose, refashioning the piles of drab gadgetry into a bona fide catalog of coming fixtures destined for well-appointed homes of the future. The examples conjured in his writings signaled an aristocratic imagination populated by butlers and maids and mansions with rooms devoted to a single, lavish use. He took to the notion of oversize computer screens eight years before the idea caught on at PARC, and his team built a lavish room to satisfy the blueprint in their minds.

Whereas PARC's office-oriented agenda had led its scientists to regard conference rooms as a natural habitat for wall-size displays, Negroponte's group opted to create a learned, leathery man cave. Their "Media Room" featured a plush Eames lounge chair, upon which a single user would rest alone in the darkness and gaze at electronic content projected onto the whole wall before him. "Effectively, it was a room-sized personal computer," wrote Steward Brand, who penned a popular book about his visits to the MIT Media Lab during the mid-'80s.[20] Enlarging the screen beyond the confines of traditional desktops afforded a spacious estate for spreading out the user's various digital assets. The project's lead designer, Richard Bolt, said the Media Room was meant to be an "informational surround."[21] For a display of such proportions, a mouse and keyboard simply would not do. The user would navigate this informational surround largely by voice. The system had been programmed to recognize a limited vocabulary of spoken phrases. Instead of leaning forward to type his commands, the user uttered them while maintaining optimal repose.

By 1984, the Media Room was able to talk back. It formulated verbal responses to a user's questions and asked for clarification when needed, which it needed constantly. MIT produced a video to demonstrate the room in action.[22] The demo opens on an architect lounging in the chair, the camera peering over his shoulder, aimed at the giant screen on the wall. The plan for the video was to show the architect having a pleasant chat with the machine; he would specify the kinds of buildings he wishes to view and the machine would display images of those buildings. He would then sift through the images one by one and ask the machine for the building's name, who built it, and so on. While the Media Room was a technical marvel for the time, even the architect in the video fails to downplay just how maddening it is to use the system. Between every moment of "awe" come many sighs of frustration. After the first minute, Negroponte and Bolt's movie-theater-for-one vision of computing devolves into an interactive self-torture chamber.

When the architect asks the machine to find images of religious buildings, the encounter begins to go awry. "Ah ha," says the machine, "There are 2 . . . 88 entries. *There* are 2 . . . 88 entries—." (It keeps repeating the answer, and its intonations vary each time.) Worried that the machine might continue forever, the architect sternly interjects, "Okay, show them!" Nothing happens on screen. Silence. "Okay," he repeats, "*show . . . them*." The machine shows them. Human and computer exchange a few choppy questions and answers about the first religious building on the screen, and the architect presses on to the next line in his script: "Next project . . . *next* project." He waits. The machine eventually replies, telling the architect to "speak up," only to execute his command a second later. Indignation creeps into the architect's voice as he begins to rock in his chair. "Who designed it?" he asks. The machine spits back: "How should I know? Why not repeat it?" The architect asks again, "Who is the designer?" The machine then reveals the name, and the demo continues for five more minutes, with escalating unease until the end.[23]

Recurrent mentions of MIT projects during Weiser's visioning meetings at PARC suggested one thing about the future: Negroponte's Media Lab would likely be a key competitor in the quest for a new kind of computer. In fact, Negroponte's earlier forays into computing pursued the same elusive goal that Weiser had sought in a series of electronic art installations he created with three friends during his time at the University of Michigan (in his spare time between graduate school, reevaluation counseling, and starting a family). Both men, unknown to each other at the time, spent the mid-1970s trying out different ways to turn ordinary environments into interactive spaces by means of electronics.

Negroponte's efforts attracted far more recognition. Blending case studies with thought experiments, he argued that a building should become more responsive to its inhabitants with the addition of computation to construction materials.[24] Outfitted in this fashion, a structure ought to alter its dimensions on the fly to custom-fit the people in its presence. An architect's job, in turn, would change. Rather than craft a polished blueprint in painstaking detail, architects could design the processes through which a room might constantly reconfigure itself according to an individual's preferences or a subset of variables the room had been programmed to monitor. Already commonplace in the skyscrapers Negroponte frequented were elevators that performed decisions about when to stop accepting new passengers. Equipped with a weight sensor and an algorithm, these attentive elevators kept a running tab on the total pounds they carried; whenever a maximum weight limit was hit, the elevator would ignore the summons of those waiting for a lift until it reached a floor where enough of its current passengers got out. Negroponte extrapolated a wealth of further possibilities: if we could program an elevator to be safety conscious, then it was just as viable to design "a courteous elevator, a suggestive elevator, a humorous elevator. . . . We can wonder about the eventuality of its being grumpy, poking fun, or trying to befriend influential passengers by giving them more personal and efficient service."[25] Beyond elevators, one could speculate wildly about outfitting whole buildings with watchful sensors and algorithmic personalities. "We can imagine the home of the future having surrogate butlers and maids embedded in all walls and floors," Negroponte wrote.[26] This notion of intelligent environments replete with electronic servants—which was central to Negroponte's work during the 1970s—was set aside during the '80s as he turned to shinier objects around the Media Lab.

Weiser's 1975 attempt at a smart room was, on the other hand, inspired by his immersion in the era's activist arts movement. Handheld camcorders such as Sony's Portapak were heralded by avant-garde youth across the country; with his Ann Arbor roommates, Weiser launched a group called Cerberus and promptly won an NEA grant to fund a series of shows they held at small museums around the Midwest.[27] They positioned screens against the gallery walls and walked around each gallery recording the crowd with a Portapak, and the video they captured would display live on the screens. "What was interesting to Mark," his wife, Vicky Reich, remembered, "was that with the Portapak and videotape and feedback, you turned a room into an interactive space. The art was people interacting with the cameras that then showed up on monitors."[28] While the results varied, the process was the essential focus. Many such artistic adventures with Portapaks were done in the spirit of Herbert Marcuse's call to expand consciousness by seizing the means of media production. Mass entertainment like 1960s television, Marcuse claimed in a book that Weiser revered, helped make individuals feel alienated.[29] Independent video experiments were seen as a breeding ground for more integrative ways of being.

And it was that sort of ambition that Weiser brought to the visioning meetings he led at PARC. Three months into these meetings now, his hopes had dimmed to a flicker as he strained to find something interesting in what his fellow computer scientists were saying.

At the opposite end of PARC's facility in its Systems Science Lab, a colleague who had befriended Weiser was pioneering the ethnographic study of modern technology users. Lucy Suchman had joined the lab as an intern in 1979 while still an anthropology doctoral student at UC Berkeley. Before that, she never imagined working in Silicon Valley. Anthropologists fascinated her as a teenager for the ways they sought to understand vastly different cultures. She was interested in Native American tribes in particular, though she realized in college that it would be better to cultivate other interests: "I was just overwhelmed by the sense of their wonderful ways of organizing and relating to the world. But it was a very terrible history, and I started to feel that the last thing Native Americans needed was another anthropologist studying them."[30] Berkeley's campus activism in the late 1960s inspired Suchman to take up a new direction in the field, which had ethnographers shifting their gaze toward American institutions and powerful companies. Shedding critical light on the everyday activities through which their power was exercised seemed like a promising avenue to spur progressive change. Suchman accepted a chance opportunity to enter Xerox's Palo Alto facility, where she would end up staying for twenty-two years.

Her curiosity about PARC's organizational dynamics soon extended to the lab's technical innovations. By the time Weiser joined PARC in 1987, Suchman's doctoral research had blossomed into a book filled with stirring implications for computing, cognitive science, and artificial intelligence.[31] Not rostered in the Computer Science Lab, she did not attend Weiser's visioning meetings. A room full of technologists musing about their designs for the future represented the antithesis of Suchman's preferred vantage point. The ordinary user they imagined in their mind's eye often held little in common with actual users. And each actual user—each person—was likely to be quite different from the next. Nonetheless, these computer scientists as a rule were quick to regard their own experiences using the prototypes they built as a stand-in for the average person's experience. Suchman looked for the details that showed up only when a device sat on someone's desk, beside a mug of coffee and photos of their kids. The ethnographic research she had performed during her first decade at PARC sought to examine office workers in their particular contexts. She parked herself right next to accountants, asking them questions and taking notes as they filled out expense reports. She sat in lots of copy rooms watching people make copies. In the details, she studied the unforeseeable factors that shaped new technologies once they had left the controlled settings of trade shows and product demonstrations.

Weiser fondly recounted later that, right around the time of his visioning meetings, Suchman and her fellow anthropologists "were criticizing computer science for its lack of relevance to life."[32] Weiser's work up to that point largely fit their critique. In civilizations throughout history, the anthropologists pointed out, technologies acquired staying power only after they became integrated into the social fabric;

fringe instruments that required people to break from the flow of their routines tended to fall out of fashion.

Historical insights and existential questions were infrequently ventured by hardware engineers in Weiser's wing of the building. Their mounting enthusiasm around computer walls was still an initiative in search of a purpose; however, it wasn't hard for computer scientists to agree that filling the world with more kinds of computers represented a valiant end in itself. Most loved computers unconditionally. Weiser did not. The computer had been his "best friend" at times, yet he often lamented the pull it had on him. The computing centers of his youth, the deep-seated ambivalence they sowed, had primed Weiser to viscerally appreciate how lines of code and graphics on screen could feel far removed from the rest of one's being.

Weiser's conversations with Suchman were occasion to step back from the technical grind. The ideas of Martin Heidegger that first captivated Weiser in his days as a philosophy major also interested Suchman, and she had shown in her book that a Silicon Valley researcher could cite esoteric German concepts to great effect. There was indeed a place—maybe even a need—for innovative theories about humanity within this high-tech citadel. Suchman's criticisms of existing technology appeared to give Weiser a more thrilling sense of computing's future than his lab's brainstorming sessions did. Suchman's book, more than Weiser's meetings, held fresh prospects. At its core was a meticulous analysis of a computing system that in retrospect could be counted among the first to gesture toward the Internet of Things, though no one used that term at the time. For her part, Suchman welcomed greater collaboration with the Computer Science Lab, so long as it was a real partnership. If technologists were willing to reckon with the experiences of actual users rather than sticking to the play spaces of their lab, then some of the recurring problems that Suchman had identified might get resolved.

Suchman's 1987 book, *Plans and Situated Actions: The Problem of Human-Machine Communication*, was inspired partly by the difficulties that arose whenever people sought to use an unfamiliar device for the first time. The enigma surrounding one Xerox machine's failures proved especially formative to her thinking. In 1981, tremors of panic spread throughout the Xerox Corporation as complaints about their latest photocopier—the 8200—came in from offices, schools, hospitals, and courtrooms. "Everybody was collapsing into a twitching heap," recalled Brown.[33] The 8200 copier was supposed to be a revolutionary product. Weighing in at 2,236 pounds and nicknamed "the chainsaw," the 8200 generated seventy copies per minute and was, as Xerox boasted, the industry's first copier to have Ethernet-enabled microcomputers built into it. "This giant (small-car-sized) machine did everything—it reduced, enlarged, sorted, collated, and even bound copies automatically," wrote the technology historian Autumn Stanley.[34] Such added functionality introduced extra levels of complexity, which could make photocopying more complicated. The company's executives had used the 8200 themselves before its release; none had experienced any of the confusion that laypeople around the world were now airing. CEO David T. Kearns was confounded by the discrepancy. Not even the product's designers could give him an explanation about what was going wrong. Desperate to understand the problem better, Kearns turned to PARC.

PARC's efforts to rescue the Xerox 8200 materialized around two fronts. The first sent Suchman (still a visiting student-researcher at the time) and the PARC scientist Austin Henderson on site visits to various workplaces, where they sat together in copy rooms watching people attempt to use the 8200 copier. Suchman wished to better understand the nature of their complaints. A few site visits with unsatisfied customers convinced her—a bit to her own surprise—that a deeper investigation was necessary. Photocopiers, no matter how advanced the model, were less complex than most electronics. But when she asked people to elaborate on the difficulties they were having, they could not put their finger on any specific cause. Hours of live observation revealed no quick fixes either. "I, as an observer of their troubles, was equally confused," Suchman reported.[35] Suchman and Henderson determined that they needed a better way to access and analyze what people were thinking about as they silently tried to make sense of the machine. Henderson conveyed their plan to Xerox executives, who agreed to fund the unconventional proposal. Rather than asking individual users to "think out loud" while using the 8200 (a common tactic then), Suchman and Henderson were hoping to employ video cameras to record pairs of users having a conversation as they attempted to make copies. A dialogue between two people seemed more authentic than a solo user trying to narrate his every thought. Cameras would be installed on the walls of a PARC room that housed an 8200 copier so that Suchman could study the encounters with the eye of a coach reviewing footage of her team's mistakes, rewinding and pausing to take in key moments.

Meanwhile, as Suchman's ethnographic study got underway, PARC's technologists hammered out a possible solution. The problem as they viewed it was rather straightforward: operating a state-of-the-art photocopier required knowledge that laypeople—the intended users for the 8200—did not wish to acquire. Richard Fikes and Dan Russell, resident specialists in artificial intelligence, had a solution: more technology. Providing lay users with an instruction manual clearly wasn't working, but what if the instructions could be programmed into the machine? What if the photocopier could communicate with the users, guiding them step-by-step through each task? Once Fikes developed the software—a program that communicated with users through a computer monitor he attached to the copier—Russell set about writing code that would furnish the copy-machine novice with a library of on-demand instructions complete with animated graphics to illustrate common tasks. They called the system Bluebonnet, a folksy name that belied all the heavy lifting involved in such a formidable technical achievement.

Toward the end of 1983, Bluebonnet was ready for testing in PARC's camera-outfitted copy room, and Suchman was keen to observe. The designers were confident that their state-of-the-art prototype would make the 8200 copier easy to use: "I knew with the hubris of a young newly minted PhD that it was a perfect system," Russell recalled.[36] For Suchman, Bluebonnet had heightened the intellectual stakes of the project; it wasn't just a copy-machine issue anymore. Fikes and Russell's software was a near-to-hand example of an ambition at the heart of AI research during its still-early years in the 1980s. Programming expert knowledge directly into the tools and machines that people used around the office promised to revolutionize the future of work. With the aid of an expert-help system, any professional—a doctor, for instance—could interface more efficiently with the vast body of knowledge from which her mastery was cultivated and maintained; the doctor might use such systems alongside her medical instruments to rapidly comb through images or descriptions of symptoms related to those exhibited by the patient on her table. Expert-help systems could also deliver relevant technical knowledge to the nonexpert in specific times of need, when he was attempting to do something he had no idea how to do. Fikes and Russell's photocopier aid strove for the latter, but something was going very wrong.

The confusion around the 8200 copier persisted and even swelled in proportion. This time around, Suchman was in a better position to make sense of it. One thing became immediately clear: you couldn't just blame the user. Anyone citing human error would have to contend with Suchman's video of Ron Kaplan and Allen Newell. When Brown visited Rochester to report on PARC's findings, he played the video for a room of Xerox managers. It featured the two men struggling and failing to produce a copy of an article Kaplan had written. Everyone back at PARC knew who Kaplan and Newell were—Kaplan was a world-renowned computational linguist, and Newell was revered as a founding father of AI. The video showed Newell peering over Kaplan's shoulder, oscillating between curiosity and confusion as the pair exhausted the full arsenal of their joint expertise, to no avail, for over an hour and a half. One of the Rochester men slammed his fist on the table and hollered, "Goddammit, Brown! What did you do, get those guys off the goddamn loading dock?" Brown looked at him and smiled, "Well, actually, let me introduce these two stooges to you."[37] He revealed Kaplan's and Newell's pedigrees, and the Xerox execs sat speechless.

Back at PARC, in order to pinpoint the shortcomings of the Bluebonnet system, Suchman dug into its underlying logic. The first oddity to pique her suspicions was the absolute confidence its creators placed in the efficacy of instructions. Such faith was at the cornerstone of "the planning model of AI"—Suchman's term for the belief, popular among AI researchers at the time, that most action humans take are derived from a corresponding plan we have stored in our heads. When we bend down to tie our shoes, for example, our fingers carry out a learned sequence for tying a knot; we don't figure it out anew each time. Suchman acknowledged that little plans like these play a crucial role in daily life, but years of scrutinizing people's work practices across various fields had her convinced that there was more going on than the planning model could account for. Bluebonnet presumed to serve as a stand-in for the plan that lay users would've had in their heads if they had taken the time to become photocopying experts (as they had done with shoe tying). Lacking any internalized sense of how to operate the machine, the users would submit to each new instruction as it was presented on Fikes and Russell's graphical display. Technologists who adhered to the planning model reasoned that a novice aided with an electronic help system would perform on a par with the experts; once you had the right plan at your disposal, it was easy to execute the correct actions.

Suchman knew that this assumption ran counter to findings in the social sciences. The task of imparting knowledge from an expert to a novice was immensely more difficult when master and pupil were not in the same room—to say nothing of the problems facing a nonhuman instructor. A body of research had recently grown up around the giving and following of basic instructions. In a dissertation devoted to the subject, Julie Burke studied a group of plumbers and their trainees as the former taught the latter how to assemble a water pump for the first time.[38] Burke asked the seasoned plumbers to instruct the trainees in multiple ways—face-to-face, over the phone, and via handwritten instructions—so that she might parse how effectively their directives worked in each case. She found that, regardless of the instructors' teaching skills, their instructions always left something out when delivered over the phone or in writing. Whenever instructors were not standing beside the trainees, they could not acknowledge or monitor the idiosyncratic circumstances that made every scenario at least a little bit unique. An instruction that worked for an average-size person might not be doable for a very short one. A caveat that was too obvious to merit mention for one trainee might have been crucial to prevent another trainee's mishap. Unless remote instructors were unusually meticulous, they tended to neglect salient details. Yet, some degree of ambiguity turned out to be necessary for giving good instructions. In order for the instructions to flow as a coherent series of discrete steps, expert plumbers needed to refrain from mentioning details that seemed less essential, so as to not drown their basic directives in a sea of qualifiers. It was better to say less and let the trainee improvise than to risk overwhelming him by saying too much. Burke's research suggested that instruction followers were constantly filling in the gaps that instruction givers needed to leave open for the sake of comprehension.

Still, the question remained: Why were aspiring plumbers able to assemble water pumps with such a lower failure rate than the computer scientists trying to make copies with the aid of Fikes and Russell's help system? Unlike a set of written instructions, Bluebonnet aspired to have a conversation with its user. But unlike a live phone call with an expert, conversations with Bluebonnet were largely one-sided. Fikes and Russell's system led each newcomer to expect that it would listen to them and reply accordingly. It would ostensibly retrieve only the information that the user appeared to need at just the moment she appeared to need it—no more flipping through printed manuals, no more trial and error. The average person would step right up to the photocopier as if asking a local resident for directions to the nearest gas station. Yet, after the system doled out its first few instructions, most everyone Suchman studied invariably felt lost.

Suchman realized that Bluebonnet had been programmed around an incomplete understanding of how conversation worked. Fikes and Russell's system assumed that conversation was simply a matter of signal transmission: person A has something to say to person B, so he says it; person B hears the message, processes it, and then retrieves an appropriate reply to utter back to person A. Messages pass between the two parties, but most of the work is done solo. When A asks B, "How's it going?" it's as if A's question gets delivered to B's attention in the form of a sealed envelope that B opens to read alone behind a closed door. Drawing on Harold Garfinkel's ethnomethodology and the conversation analysts Charles and Marjorie Goodwin's studies of everyday spoken exchanges, Suchman insisted that ordinary conversations were rarely such divisible, linear transactions: "Closer analysis of face-to-face communication indicate that conversation is not so much an alternating series of actions and reactions between individuals as it is a joint action accomplished through the participants' continuous engagement in speaking and listening."[39] Conversations were a collaboration. If anything, Bluebonnet more closely resembled a scripted survey interview in which "the choice and order of topics—what is talked about and when—is established by an absent third party."[40] The programmer bequeathed a plan to the system, and the system lacked the capacity to recognize even the slightest deviation from that plan. As far as the system was concerned, either a user's action corresponded to some step in the programmer's plan or that action did not exist. The system could only guess at the user's action within the framework of the programmer's intentions. Users were led to regard the machine as an attentive coach who was alert to the meaning of their every move, but Bluebonnet couldn't actually hear, see, or feel much of what they were doing to the machine.

Ultimately, Suchman concluded that the false promise of interactivity that was being levied onto the 8200 copier and other so-called intelligent, responsive machines was unwittingly courting users and devices into an *asymmetrical* relationship devoid of genuine conversational dynamics. Human interaction was typically symmetrical in that each party enjoyed access to a similar set of capacities for making sense of a shared moment. Whenever two humans conversed face-to-face, each generally drew upon a range of communicative resources extending well beyond what most machines could compute. Auditory individuals surmised meaning from the slightest shift in vocal tones. Sighted people noticed anything they felt compelled to look at. Expert-help systems could register only the absence or presence of documents and bodies that happened to pass over their sensors and press their buttons. To fully dispel the notion that Bluebonnet's encounters with its users were anything like a conversation, Suchman transcribed the exchanges she captured on video using a four-column chart, which accompanied her book's lengthy analysis of the help system's woes. In the chart's first column, she kept a running record of everything the pair of users said aloud and gestured at while attempting to make their copies. The second column documented actions they took that were actually intelligible to the help system—actions such as opening or closing one of the copier's components, inserting a document into one of its trays, or pressing the start button. Entries in the first column (utterances and gestures that were unintelligible to the system) vastly outnumbered those in the second column. The users consistently wanted to engage with the system in ways that it could not process, as evidenced by the entries dotting the third column in Suchman's chart. Indicated in the third column were the system's displayed instructions—the display changed only when the user performed a second-column action (that is, an action Bluebonnet recognized). The chart's fourth column summarized the rationale associated with each change in the display. When this fourth column was compared against the users' corresponding first-column activities, it was clear that the users didn't understand the system's rationale. More often than not, whenever the system's display changed to present a new instruction, the users reacted by talking among themselves in an effort to figure out why the display had changed.[41]

Just as the users were generally left to guess at the meaning of Bluebonnet's instructions—and its inactivity in the face of their expectations for further guidance—the system itself was constantly having to make crude assumptions about the few user actions it was equipped to detect. For instance, while Bluebonnet could register that a document had been placed in or removed from the copier's "recirculating document holder" (RDH), it could not decipher whether the user had inserted the entirety of their document or just the first page. Thus, as Suchman noted, "the system just takes the evidence that *something* has been put into the RDH as an appropriate response, and takes whatever is put there as satisfying [its instructions]."[42] If the user did not in fact place the entirety of their document in the RDH, however, Bluebonnet had no way of detecting this. It simply proceeded to displaying the next step, under the assumption that all the user's pages had been copied; only the user knew that there were more pages remaining. During these moments, users often stood silent for quite a while, just waiting for the help system to walk them through the rest of their pages. The system, for its part, was waiting for the user to move on. "It was as if the machine was looking through this tiny little pinhole at what was going on," Suchman said.[43] Whenever an impasse occurred, the system either sat inactive or repeated a prior instruction, leaving the user to wonder if it was even functioning: Was the system really asking them to repeat something, or was it just frozen? Either way, until the user had completed the present step in precisely the way the system could recognize, the system went on waiting, as it could neither rephrase the instruction nor stop the user from taking an erroneous action. This "insensitivity to particulars of the user's situation," Suchman argued, was "the limiting factor on [Bluebonnet's] ability to assess the significance of her actions."[44]

Suchman contended most computer interfaces, be they expert systems like Bluebonnet or conventional software running on a desktop, suffered from a limited capacity to access, let alone understand, "the moment-by-moment contingencies that constitute the conditions of situated interaction."[45] The continuous stream of sensory details that we live by—the tacit dimension, in Polanyi's verbiage—was just a bunch of white noise to even the most sophisticated programs. The disparity between humans and machines in this critical area lay at the bottom of most user complaints. Human-machine communication, even in cutting-edge labs at PARC, was a wildly asymmetrical affair.

One might blame that asymmetry on the era's technical limitations, but those limitations could also be traced to the planning model of AI that most technologists sought to advance. They believed that the best way to help people was to equip machines with elaborate maps of all the moves one should make to accomplish a desired result. Software operating within a rectangular screen furnished a clean slate of cognition through which AI was supposed to guide users to new heights of productivity. So much attention had been bound up in the maps, so little given to the territory.

When Suchman spoke to Weiser about computer science's lack of relevance to life, she was referring to the asymmetries at play in human-computer encounters. From her remarks, Weiser pieced together his own diagnosis of the problem, but it didn't exactly match Suchman's sense of the problem—they would each gravitate toward different solutions. If, Weiser appeared to reason, personal computers were, like Bluebonnet, stymied by their "insensitivity to particulars of the user's situation," then perhaps computer science could correct its course by figuring out how to make computation more attuned to people. Attaching Bluebonnet to the 8200 copier failed to strike up meaningful human-computer exchanges, though it was still an interesting attempt to embed computing in a daily activity away from one's desk. Perhaps there were other devices he could attach to other objects around the office, but in a manner that rectified the shortcomings Suchman had pinpointed. This ambition seemed to grow in the recollected afterglow of his conversations with Suchman. Whenever Weiser later recounted the origins of "ubiquitous computing," he would usually cite Suchman as a key influence, and she wouldn't quite know what to make of that. Whereas Weiser would make it his mission to imagine new kinds of computers, taking PARC's rooms as his test bed, Suchman went on to cultivate alternative approaches to technical innovation that ventured outside the lab. For her, the insensitivity at play in certain technologies had more to do with the self-referential processes that steered their design.

Inspired by his interpretation of Suchman's analysis and its perceived reverberations with philosophical notions from his youth,

Weiser began plotting a new line of questions to bring back to his fellow computer scientists, which he later summarized: "How were computers embedded within the complex social framework of daily activities? And how did they interplay with the rest of our densely woven physical environment (also known as 'the real world')?"[46] Anthropologists saw in such questions years' worth of careful observations to conduct in many disparate settings. To them, "physical environment," "daily activities," and the like were just placeholder phrases that mandated serious fieldwork in the trenches at specific locales: actual copy rooms, classrooms, law offices, and so on. Weiser tended to entertain these questions on a theoretical plane where they applied to everywhere and nowhere in particular. You could arrive at answers quicker that way, and the answers might sound grander. If you asked yourself how desktop PCs interplayed with the real world (the world as you experienced it to be), then the conclusion was obvious—at least, it was in Weiser's case. PCs "colonized" your life, he would eventually insist.[47]

Decades later, when reflecting on her first chats with Weiser, Suchman would say, "He had a very romantic relationship with anthropology."[48] It wasn't a compliment. She was never really convinced, not even in those early years, that computers would ever converse with us in any genuine sense. The asymmetries barring machines from the subtle cues of human communication were not a trivial matter. They could hardly be compensated for by great technical leaps. Robotics and AI raised fascinating prospects in their own right, but they appeared ill suited to replicate the sense of mutual understanding we get from a wry smile, a perfectly timed wink, or just gazing into another's eyes. Installing sensors and software into everyday settings might make computing more convenient, but doing so could also make human-computer asymmetries a ubiquitous aspect of high-tech existence.

The days leading up to Halloween frightened a thinning network of Xerox diplomats who kept PARC researchers tied to the company's executives back east. As they took their kids costume shopping and drove past skeletons and ghosts cropping up on neighborhood lawns, they must've braced themselves for October 25, 1988—the day bookstore shelves would display *Fumbling the Future: How Xerox Invented, Then Ignored, the First Personal Computer*. The book had been years in the making and promised readers an inside look at one of the biggest missed opportunities in business history. By and large, its authors, Douglas Smith and Robert Alexander, cast PARC's computer scientists as noble, aloof geniuses whose breakthrough devices—chiefly the Alto desktop and its graphical user interface—flew over the heads of Xerox's business leadership. *Fumbling the Future* was not the first piece of writing to expose the cognitive dissonance and mutual suspensions that frayed relations between the two cultures at the PARC campus and at Xerox Tower, the corporation's massive Rochester headquarters. But it was the first book, and a major one that garnered waves of publicity. Smith and Alexander contended that the in-house tug-of-war over personal computing "smoldered unproductively within the company for more than a decade."[49] Certifying the book's damning depiction of Xerox's Rochester executives, the *Economist* concluded: "If Xerox had listened to its researchers, it stood to win a bonanza even bigger than copiers had produced. In the first decade of personal computing, the authors reckon that revenues amounted to $10 billion."[50] To the battle-scarred peacemakers who chose to stay at PARC and at Xerox after having lived through those tense episodes in the late 1970s and early '80s, every headline about the new book seemed to make their present jobs a little bit harder.

However, most researchers in the Computer Science Lab (which Weiser had just been tapped to manage) loved the book. They felt that its authors clearly stated the facts of their case against Xerox management and succeeded in showing the public that PARC—not Apple, not IBM, not Microsoft—was the true cradle of personal computing. This view already held sway within their lab. And here they were in the same space, the same terraces and meeting rooms, from which talk of the graphical user interface and Ethernet hardened into technical realities that were now sweeping the world. The tradition was theirs to carry forward.

Weiser's reflections on this legacy set him in a different direction than most of his staff. He observed that the bulk of Silicon Valley was bent on making incremental advancements to the big ideas that PARC's Alto PC had introduced. "Its highest ideal," Weiser would say of the industry, "is to make a computer so exciting, so wonderful, so interesting, that we never want to be without it."[51]

What Weiser found most riveting in the current buzz about PARC's golden decade was his predecessors' audacity to challenge the conviction that mainframe-bound, time-sharing terminals were rendering other form factors unnecessary. *Fumbling the Future* brought to light the fact that a mainframe-oriented faction of Xerox fought against the proposal to launch PARC in 1970. Those objections came from the heads of Scientific Data Systems, a mainframe computer outfit that Xerox had just spent $900 million acquiring in hopes of competing with IBM. The SDS people argued, unsuccessfully, that the investment in a center for basic research would be more wisely spent on beefing up SDS's existing product line. Xerox leadership held firm, countering that "building another mainframe computer, even a bigger, faster, more powerful one, was not in the plan."[52] There was a lesson in that opening schism, and it suggested another way of understanding one's charge as a PARC computer scientist. The center had become synonymous with personal computing because its researchers invented the building blocks that initiated a new paradigm. Those radical acts of risky invention, more so than any single innovation they produced, had made PARC what it was. To merely refine personal computers and their components—to go the way of bigger, faster, more powerful—would be, under this interpretation of PARC's virtue, to dishonor its founding spirit. While Weiser had spent months trying to nudge his colleagues to imagine what devices might come after desktops, he began November 1988 with a greater sense of urgency. PARC was due for a renaissance. If his lab was to continue inventing the future, Weiser had to compel his colleagues to let go of the technology they'd inherited.

Something else happened to Weiser during those months: his father, David, passed away. Little by little, they had fallen more out of touch in recent years. Weiser no longer made a point of sending his father letters filled with updates about his work. Most of those letters had gone unanswered; for every several Weiser sent, his father might dash off a single hurried reply. Still, even as the correspondence waned, Weiser's life kept looping back to his father's. Becoming a professor and a scientist, the time spent on a computer figuring out some new puzzle, the persistent allure of ideas from books left open around their old Stony Brook living room. Weiser had gained a lot from following his father's interests. He had built their bonds upon the faintest connections. Maybe, probably, perhaps—one always needed to infer—his father would be proud of what he was up to now.

Weiser held his lab's final visioning meeting that November at a nearby aquarium. PARC had a tradition of scheduling important sessions in quirky off-site locales. Weiser, though, devoted far less energy to preparing his notes for this last meeting than he had for the earlier ones. This daylong retreat would have been, according to the initial plan, an exhilarating occasion for the staff to connect the dots of their prior discussion, stand back, and behold a treasure map of innovations they might cocreate over the next two decades. Instead, Weiser just read aloud through a mess of bullet points listing the various stuff his peers had mentioned.[53] He could see that nothing would come of it, and he was ready to move on. Years later, when a reporter asked him about it, he laughed about the whole visioning-meeting process and called it "a disaster."[54] But there were two moments toward the end of this last meeting, fruitless as it seemed then, that did foreshadow the new agenda they would soon agree to pursue.

Sitting in the aquarium, as the staff gazed at one another in the absence of a unifying vision, Weiser finally blurted out an idea that had consumed him during a recent morning shower. He had while showering felt the urge to write down his thoughts, right away, before they escaped him. Then it occurred to him, he said, that he could design a cheap, small computer for exactly that purpose. Soon his mind cast about for more activities in other settings that might benefit from having small doses of computation within reach, just in case a need arose. Small, one-feature computers could basically be installed all over a home: in the kitchen, in the shower, on the nightstand. And all throughout office buildings, too: attached to the doors and along the hallways, in pens, and on whiteboards.

Weiser scanned his colleagues' faces for signs of support. "Everyone thought it was the worst idea they had ever heard," he recalled. "Some people said, 'That would be boring' or 'That would be too easy.' But I persisted."[55]

Smallness was a virtue in Silicon Valley, but only in the sense of shrinking electronic components so as to pack more features inside a personal workstation. You made things smaller so that you could make them more powerful. You generally didn't miniaturize computers just for the sake of mobility or simplicity; little gadgets were still relegated to the margins of popular computing at that point. Desktop PCs were celebrated because they stuffed an array of features and programs into one box, giving the user the ability to do seemingly everything all at once and to the max, without having to leave her seat. Weiser's shower epiphany sounded like "Chopsticks" to the ears of classically trained technologists. "At the time it was very counterintuitive," said Paul Saffo, Weiser's friend and the former director of the Institute for the Future. "Everyone back then was agglomerating more applications for the one device you had in your life, the PC."[56] Most hardware and software engineers working on improvements to the PC believed the machine to be the field's surest path of ascension toward "man-computer symbiosis" and other lofty ideals. It was as if Weiser, with his strange metaphors and trifling anecdotes, was now trying to beckon them down the path into some unnamed jungle.

After Weiser's shower story, the aquarium meeting ticked down its agenda to a reading Weiser had included in the pre-meeting materials. Entitled "TABLET: Personal Computing the Year 2000," the article showcased a team of students from the University of Illinois who had just won a contest at Apple headquarters. Middle-age computer professionals were often peering over the work of twentysomethings, hoping to find something they might nurture for themselves. In this case, the

Illinois students' product proposal was remarkably sophisticated: they outlined in clear prose their design for an electronic tablet the size of a paper notepad. Embracing the analogy, they cited paper's light weight and portability as key properties that future computers ought to mimic. "We seek something that will fit comfortably into people's lives. . . . No one thinks twice about taking a pad anywhere," they noted approvingly before listing their digital pad's specifications.[57] To Weiser, this passing comparison with paper artifacts was the article's most valuable bit. He saw how couching new ideas in such terms might act as currency around Xerox, a company that wished to develop the electronic frontier without displacing paper in the process. (Much of their revenue still came from copying, printing, ink, and toner.)

More immediately, the students' design provided a more solid ground for discussing his nebulous goals. The wireless tablet offered an initial form factor for bringing the machine out into the world. Going from a desktop to a tablet at least got the machine moving. Wherever people went, portable computers could be carried along easily. Apple's endorsement lent some credibility to the notion, especially because the contest's judges included not only Apple cofounder Steve Wozniak and the sci-fi legend Ray Bradbury, but most notably Alan Kay. Though Kay no longer worked at PARC, his contributions to the Xerox Alto had earned him royal stature around the center's Computer Science Lab—his blessing was not something to scoff at. Kay's fondness for the Illinois tablet made imminent sense. The students' design was essentially an updated version of his 1972 "Dynabook" concept infused with 1980s technology. Their proposed tablet would feature an LCD touch-sensitive display; a LaserCard for data storage; a built-in microphone, audio speaker, and GPS receiver; and even an infrared bar that wrapped around its edges. This last feature left an impression on Weiser. The "IR bar" would enable the tablet to "connect to its immediate surroundings," the students wrote, as it would allow data to pass wirelessly from the device to nearby objects, including "printers and projectors, stereo headsets and video cameras, toasters and roasters, and just about anything else."[58] The students' notion of establishing proximate connectiv-

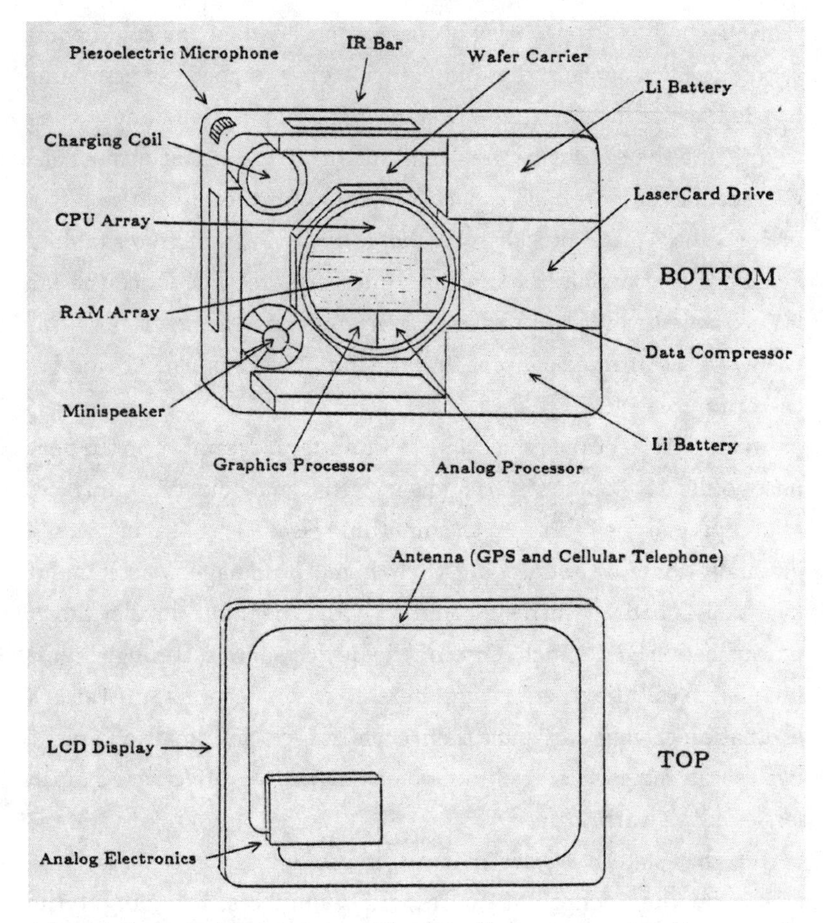

A diagram illustrating key components of the 1988 tablet computer proposed by a team of students at the University of Illinois. Courtesy of NASA.

ity with such things wasn't their primary focus, yet it hinted at a different side of portable computing than Kay had originally. In any case, tablets—whatever designs they might take—were starting to become more technologically viable. Any computer scientist familiar with Kay's work (as everyone in Weiser's lab was) could recall that Kay regarded the Alto workstation, then and now, as an "Interim Dynabook." It was a prospect they needed to consider.

The tablet served Weiser well as a point of leverage in the months after that final visioning meeting. Inspiring hardware engineers to move beyond desktop workstations was easier when you could point to some gadget at the end of the tunnel. Weiser ditched the first name he tried to give the initiative—"scrap computers"—as nobody wanted to imagine the output of their craftsmanship to be that disposable. CSL settled on more self-respectable jargon for their in-house tablet project. The "Pslate" quickly took shape, mirroring the Illinois tablet's keyboard-free touchscreen and its fine-tipped stylus. Once the lab hacked together a basic prototype, Weiser planned to experiment further with the infrared concept that the Illinois students had floated in their proposal.

In fact, Weiser called the Pslate "a short-term dream" from the very onset of its development.[59] He now viewed Dynabook-style interfaces as stepping-stones en route to something else. His colleagues would gradually come to understand that their lab manager's excitement over tablets had very little in common with other Silicon Valley tablet enthusiasts. His distinct sense of the device showed through in his impulse to call them scrap computers. Each single tablet, in Weiser's estimation of what was viable then, could at best "do a small job well."[60] Each single tablet, if left to function on its own, was effectively a much less effective workstation.

But the Pslate as Weiser imagined it would not be another isolated computer. The strength of Pslates was in numbers, according to the new pitch Weiser had been delivering to one colleague after another, in their offices or at the coffeepot. Setting aside the capacities of one Pslate, the real question—the exciting challenge—was what a group of users could do with a fleet of Pslates that all talked to one another. The average office worker, Weiser insisted, would someday own five to ten Pslates, and there might be upward of a thousand present in buildings the size of PARC. Calling for that many screens was a baffling proposition to even the most devout techies. What, for instance, would a team of six people do with sixty tablets? Wouldn't they be better off with six state-of-the-art workstations? What could you do with sixty quasi-computers that couldn't be done with six powerful desktops? Weiser didn't have anything particular in mind; examples were not his forte, and they would come later. These prototypes-to-come would, arguably more than anything else built during the 1990s, of course serve as important precursors for twenty-first-century iPhones and iPads, even though Weiser himself would gradually come to favor devices that no longer revolved around screens.

For now, screens remained central, and he concentrated on his broader thesis: if computers were everywhere—some big and many small, all continuously connected—then using a computer would no longer need to be such an isolating experience. Computing might start to become more integrated, more relevant to life. Whereas most technologists dreamt about the applications their prototypes could soon enable, Weiser's reflections about new systems were primarily concerned with the worldview that they might help foster. He had circled an upcoming day on his schedule: May 15, 1989—the day he would unveil his budding vision to the rest of Silicon Valley, not with a demo, but through an elaborate chain of concepts.

4

Tabs, Pads, and Boards

EVENTS HOSTED BY THE MIT CLUB of Northern California weren't like the ones normally put on by other alumni groups. Members gathered together without sports, beer kegs, or makeshift mascots. Anchored in Silicon Valley, the club convened around talks delivered by noteworthy local figures. Most people in the club worked in the tech scene, as did many of the speakers they invited. This was true on May 15, 1989, when Mark Weiser spoke on "Ubiquitous Computing."

The *New York Times* had done Weiser a great favor. A day before his talk, the paper ran a story titled "The Big News in Tiny Computers" by its Bay Area correspondent, John Markoff. The article profiled a handful of nearby startups that were each preparing to sell a handheld, pocket-size PC. "We're talking about a new device for a new kind of computer user," declared S. Jerrold Kaplan, a cofounder of a startup called GO. Markoff corroborated the hype, citing explosive sales figures in portable machines during recent months and suggesting that the next breed of even smaller interfaces could make for a multibillion-dollar market.[1] The MIT Club had thus shown up at Weiser's talk hungry for an expert's assessment of this lucrative frontier. It was still anyone's game, according to Markoff, as big names like IBM and Apple were then just tinkering with prototypes behind closed doors. "No one is really certain what these new go-with-you computers will do," Markoff reported.[2] Well-funded upstarts like GO and Poqet hypothesized that American electronics users longed to enjoy continuous access to digital information wherever they went. The soon-to-be-launched

The Poqet PC handheld computer. Photograph by Michael Hicks.

"Poqet PC," roughly the size of a VHS tape, featured word-processing and spreadsheet software on a screen capable of displaying twenty-five lines of text. With a fully functional little keyboard, the device looked much like a laptop, and, running for up to a hundred hours on two AA batteries, it proved more power efficient than machines three times its size. Poqet PC users could also share files remotely with their desktops by hooking up the handheld to a portable dial-up modem. Equipping workers with anytime-anywhere computer access would boost productivity, Poqet's ads promised.

Go's handheld, still early in its development at the time, would abandon the keyboard in favor of a stylus, betting that the ability to handwrite on the screen would appeal to businesspeople who increasingly relied on electronic documents, even as they delegated typing and other computing-heavy tasks to their secretaries. Handhelds would give them an easy way to check in and watch over things. But these selling points stood on the thin ice of novelty, especially when weighed against their price—the Poqet PC would go on sale that October for $1,995 (roughly equivalent to $4,500 in 2022).

Tiny computers boasted a small suite of desktop-like applications, yet no one had made a case for their larger significance. While emerging handhelds like the Poqet PC were capturing the industry's imagination, the assumptions underlying their designs were antithetical to the future Weiser wished to bring about. To his thinking, the gadgetry underlying tiny computers offered a revolutionary means to *break away from personal computers* in search of a better model. Such was the message he had arrived eager to deliver in his presentation at the MIT Club.

As Weiser slid his first slide onto the projector, a puzzled hush spread from the front row to those standing at the back of the room. Gazes lingered on the analogy Weiser had put on display. "If you remember nothing else . . ." his slide read, "Personal Computers are to Ubiquitous Computing as Strip Mining is to Ecology."[3] After a moment, Weiser flipped to the next slide and began his talk. He'd revisit strip mining in a bit.

First, he wanted to explain that new term couched in his analogy: *ubiquitous computing*. It was common parlance among the tech set to say that computers were becoming ubiquitous throughout society. Weiser had something more in mind. He coined the phrase to signal not just an increase but a total overhaul. But before mentioning the devices his lab was working on at PARC, he felt it imperative to describe, in some detail, the fundamental belief that had inspired his vision.

"Ubiquitous computing," said Weiser, "starts from a new view of people."[4] It was new to this crowd, anyway. Without naming the philosophers from his youth, Weiser outlined how all manner of human activity, from breathing to thinking, was the outcome of a tacit, running dialogue between the body and its surroundings. To speak of a human as an individual unit of life, as Western thinkers had been doing for centuries, was to misconstrue our basic operating system. Weiser alluded instead to the ecological perspective then associated with Gregory Bateson, the famed anthropologist whom Weiser and Reich had taken a cross-country road trip to meet in their twenties. "The basic unit of survival is a flexible organism-in-its-environment," Bateson had claimed.[5] Now riffing off Bateson, Weiser insisted: "We are not separate from, but are inseparably reliant on, a world around us."[6] He went so far as to supply the audience with a new word: "horld," meant to denote this inseparability of *human* and *world*.[7]

The question of how to regard people—how to understand our relationship with the world—lay at the crux of Weiser's distinction between ubiquitous computing and personal computers. The former, as his opening analogy had stipulated, was ecological in precisely Bateson's sense of ecology; the latter was like strip mining. Ubiquitous computing, in keeping with its relational ontology, would "strive for a maximal mixing of humans and computation," said Weiser, which was really to say that his approach aimed to dramatically extend computing into the multitude of environments in which humans existed.[8] The mixing of humans and computers, from mainframe computing centers to the Poqet PC, had until this point largely mirrored a Cartesian view of people, Weiser argued. He contended that personal computers—even the handheld models—effectively "[broke] down the human/world system," because PCs were "built as though humans and world were separate."[9] From J. C. R. Licklider's "man-computer symbiosis" ideal to Steve Jobs's "bicycle for our minds" metaphor, PCs fashioned arrays of feedback loops that sought to amplify the cognitive abilities of a stationary, attentive human with the machine's alien powers. This notion that sitting people down in front of a general-purpose, interactive screen was the best way to optimize their mental performance had, Weiser indicated, unwittingly miscalculated how much everyone stood to lose in the bargain. The strength of new capabilities born from this pairing had led computer scientists to focus exclusively on user-computer feedback loops at the expense of all the other cybernetic ties that coupled humans with the broader physical environment.

To help visualize these assertions, Weiser showed the audience two slides he had drawn. Under the heading "Current Computing Technology," he had drawn a thick, dark rectangle meant to signify

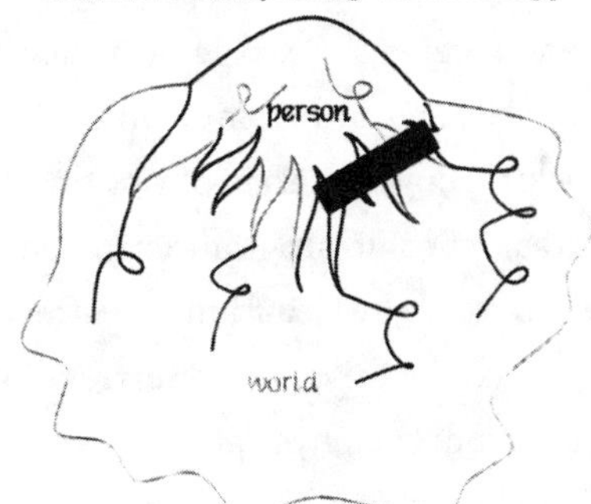

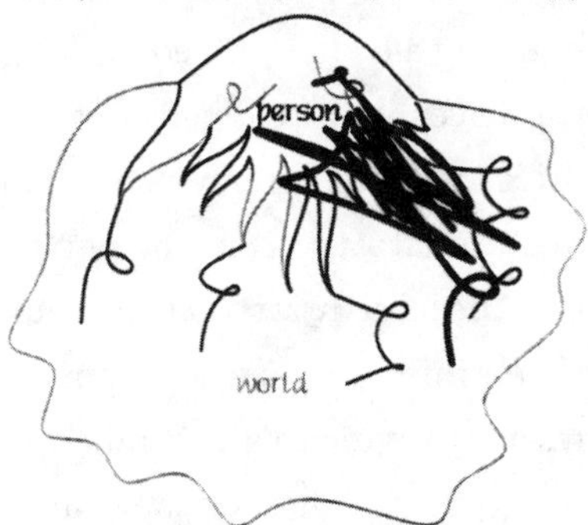

Slides from Mark Weiser's presentation on ubiquitous computing at the MIT Club of Northern California, May 15, 1989. Courtesy of the Department of Special Collections, Stanford University Libraries, Stanford, California.

a PC monitor. Its off-center placement appeared intentionally arbitrary, as it partially obscured "person" from "world" and indicated no traces of connection with the wider world beyond. The other slide, "Ubiquitous Computing Technology," looked decidedly messier and ill defined, though it conveyed Weiser's ambition to make computing more amenable to the dynamic, ecological "horld."[10] Whereas personal computing brought to mind images of an individual gazing into a single screen of his own, Weiser's notion of ubiquitous computing involved many species of digital machines that were subtly scattered across human habitats. People living in a ubiquitous computing scenario would less often use desktops, laptops, or Poqet PCs to access software, digital data, and electronic files. Instead, most of the time, they would just go on using familiar objects as they always had—writing notes with a pen and notepad, say, or driving their car to work—only those objects then would have bits of computing built into them.

Each object would feature software relevant to the task people generally used that object for. A pen and notepad, for instance, might be imbued with word-processing functions like copy-and-paste or a dictionary program, while a car might feature GPS navigation. Unlike the PC, no single object would contain a multitude of programs. Each object, however, would be designed to share data and connect with other objects embedded in the vicinity. A computerized car might, upon entering a computerized parking garage, receive a message notifying the driver about an open spot on the third floor. At no point would the driver need to type in a command, click through a series of options, or stare into a screen. The network of task-specific programs stitched into the local infrastructure would work to read the situation and offer timely information where appropriate, which the driver could reference without having to divert much attention from the physical scene. Bits of digital guidance, be they audio or visual cues, would grace the driver's awareness like an intuition—all without, in this case, gathering or storing or selling any personal data about the driver.

That was a far-off sketch of how ubiquitous computing might someday operate ecologically in sync with humans as they moved through the world. Even the simplest instance of ubiquitous computing would, Weiser knew, first required major advances in wireless connectivity, sensor networks, information architecture, and battery life. And those were just the technical barriers. The truly daunting feats, he told the MIT Club, would be the "conceptual breakthroughs."[11] Nevertheless, he was pleased to report that the first leg of the journey was underway.

He and his colleagues had recently begun work on three stepping-stones that just might transform PARC into a prescient showroom of technologies to come. Soon they would begin making ubiquitous computing prototypes in "tiny," "medium," and "large" varieties.[12] The tiny ones would fit in the palm of an adult's hand. The medium devices, like the Pslate tablet they created, featured a screen the size of a paper notebook. The large ones were comparable to a bookshelf or big-screen television. None of these prototypes would look all that different from the new handhelds and tablet PCs in development elsewhere, but they would be designed to do things that no PC could ever do.

Weiser then put his first slide back onto the projector. He still needed to explain the remaining part of his opening analogy—strip mining. Again, the audience stared up at the slide: "If you remember nothing else . . . Personal Computers are to Ubiquitous Computing as

Strip Mining is to Ecology." The connection was still a mystery to anyone who hadn't read Martin Heidegger's essay "The Question Concerning Technology." Weiser had read it many times, and his portrayal of PCs throughout his talk had much in common, not coincidentally, with Heidegger's descriptions of coal mining and other resource-extraction practices.

The essay featured a string of examples having to do with energy production. Heidegger, writing in 1954, believed that the basic existential bend of a given society was forged by how it went about deriving energy from nature. He suggested that mining coal to generate power, as opposed to using more traditional mills to channel flows of wind and water, had initiated a profound and lamentable break from earlier modes of being. Whereas older energy systems gently harnessed the natural circulation of water and wind, industrial-age strip mining imposed human order upon the land to an unprecedented degree. The enculturated meaning of landscapes slowly changed, first in the eyes of pioneering industrialists and then for miners as well as consumers. Heidegger argued that such extraction processes, which power modern societies, instilled a tendency for people to view nature as "a standing reserve"—a vast reservoir of raw materials whose value and identity is defined according to a narrow subset of anthropocentric interests (profit, heat, comfort, convenience, pleasure, etc.). Dams were another of Heidegger's pet monstrosities. The erection of a hydroelectric plant transformed his beloved Rhine River (or any river) into "a water-power supplier."[13] The water rushes on still, but our relationship to the river has changed, Heidegger insisted; it has become subject to management and remade into a kind of gear in the sprawling machinery that produces the electricity and combustions fueling modern economies.

Heidegger was not exactly an environmentalist; his ultimate lament, however, was that seeing landscapes as supply heaps drained them of their poetic power. The sheer existence of a dam or a mine levied commodity associations over even the most distant wilderness. Even a natural wonder in a national park became "an object on call for inspection by a tour group ordered there by the vacation industry," its grandeur tainted by the ominous air of engineered interventions and financial incentives.[14]

Weiser's analogy and his accompanying explanation posited that this legacy lived on in the machines that he and his audience had been making—personal computers, he maintained, were the digital-age equivalent of strip mining. The analogy hinged squarely on the parallel Weiser saw between Heidegger's depiction of industrial energy systems and the comparable manner in which the PC seemed to nag and gnaw at the corners of our minds, perpetually bending our attention to favor the electronically refined essence of things animated as pure pixels—filtered, extracted, stored, and ready to be tapped. Once transcoded into graphics, trash cans and file folders operated as icons whose only apparent grounding in the material world was a baseline semblance to physical items around the office. Establishing a visual likeness that users recognized at a glance freed the software engineer to imbue virtual icons with powers far beyond their physical counterparts. The trash-can icon could hold much more than any metal one, and it was far easier to empty or to retrieve items placed in it by mistake. A familiar item's essence was extracted and enhanced through digital modifications. Graphical simulations accumulated on the PC through a process that plucked everyday objects from their local contexts and, in turn, beckoned users to forget about the off-screen world for a while.

Personal computers, Weiser would soon write, were introducing "an unhealthy centripetal force . . . into life and the workplace."[15] For someone who lamented the pull that card punch machines and timesharing terminals had on him in 1968, it was not surprising that Weiser would be especially sensitive to and alarmed by the PC's increasing attentional allure. Two decades spent computing on devices of growing power had Weiser concerned about the existential consequences of a looming, society-scale reallocation of attention that might play out in the decades to come. If one reasoned that PCs were like mines and dams, insofar as they extracted content and attention from the rest of life, then it was hard not to worry that the undigitized varieties of worldly experience might suffer a fate similar to Heidegger's river. One's life apart from the screen might dwindle into a standing reserve for one's on-screen activities. The visceral stuff of Polanyi's tacit dimension could become merely a finicky database of sensory details to be extracted, managed, and circulated in the service of digital interactions.

By lumping PCs in with coal mines, Weiser was not suggesting that the computer scientists in the audience ought to swear off digital technologies altogether. Personal workstations—be they desktops or handhelds—distanced people and things from their traditional environments, but computing need not remain wedded to such forms. It did not have to traffic only in on-screen simulations of the physical world. By virtue of the possibilities then raised by smaller components and wireless connectivity, computing might feasibly and fruitfully establish a concrete presence amid the social fabric of everyday life. The future that Weiser left the MIT Club to ponder was a future in which "every part of your life uses computation."[16]

Those closing words gave way to applause that lasted longer than usual. It was the first moment that seemed like a real turning point since Weiser arrived in Silicon Valley. With this term, *ubiquitous computing*, he had found a perch on which he could stand above the present frenzy for PCs and squint at the twenty-first century with unique authority. Allusions to PARC's triplet of screens were already showing up in the industry's periodicals. A few months before Weiser's talk, the president of Apple products, Jean-Louis Gassee, told *InfoWorld* readers that computers would by the year 2000 "assume a much broader range of shapes," and he listed three sample shapes identical to those Weiser had in the works.[17] Gassee went on, however, to say that these new form factors represented "the second age of personal computing." Such was the initial interpretation, even among tech leaders who had been following the early buzz around Weiser's ideas. The last thing Weiser wanted was to replicate desktop workstations; yet that was the first prospect most technologists saw. Almost as soon as Weiser shared his vision beyond PARC's walls, industry insiders touted its promise without realizing they were mischaracterizing his fundamental ambition.

But those in the audience at Weiser's talk that evening must've realized that ubiquitous computing was not aiming to extend the status quo. Weiser's contempt for PCs was palpable. More melancholic than angry, his tone resembled that of a man in confession when he lambasted the machine.

By the fall of 1989, Weiser was running low on new things to say about ubiquitous computing; but what he had said thus far proved to be enough. The more attention his message attracted elsewhere, the more buy-in it received around PARC. Pictures of Weiser appeared in three magazine articles profiling his big idea and quoting him at length. *InfoWorld* printed one of his sayings in big, bold letters: "We want to make using a computer more like a walk in the woods than a visit to the dentist office."[18] *Benchmark*, a Xerox publication spotlighting research around the company, forecasted that "the personal computer, as an object on a desk, may become a thing of history."[19] *PC World* wrote that Weiser's imagined devices stood to become "so ingrained in our lifestyle that using them [would be] as natural as putting on clothes."[20] The journalists likely all found out about Weiser from the same source. Bob Metcalfe, a PARC legend who invented Ethernet before leaving to start 3Com, had said in a recent interview, "My recommendation of a look at computing in the year 2000 is to visit PARC today."[21] You couldn't visit PARC then without bumping into one of the ubiquitous computing projects that were underway. Some insiders now called the initiative "ubicomp," as if it were already a thriving movement that everyone knew about. This shorthand name stuck.

The enthusiasm mounting around Weiser had thrown him into a bind. He knew the road he did not want technology to head down. He had the industry's ear and Xerox's support. Brown was particularly bullish about the potential of Weiser's vision to inspire breakthrough devices and unify PARC. It was imperative for Weiser to leverage his recent momentum, Brown insisted: "Your challenge in the upcoming year will be to get more than poetic buy-in."[22] In Weiser's annual performance review, Brown encouraged him to be "more aggressively on ubiquitous computing" and to get "a series of projects launched."[23] And yet, Weiser found himself unsure about how to proceed beyond the figurative language and grand promises that punctuated his talks. His on-stage successes—his hints at PARC's plan to create a network of "tiny," "medium," and "large" ubicomp interfaces—pinned him up against a question he struggled to answer back in his office: "Could I design a radically new kind of computer that could more deeply participate in the world of people?"[24] He realized he couldn't—at least, not yet. Weiser recalled the pickle in writing a few years later: "As I began to glimpse what such an information appliance might look like, I saw that it would be so different from today's computer that I could not begin to understand or build it."[25] Still, he knew he and his colleagues in the Computer Science Lab needed to hack something together.

Nine teams of CSL scientists had formed suddenly around ubiquitous computing, and they were ready to start making prototypes. Ideas in the lab were either quickly built or quietly discarded. This time-honored sense of urgency with which lab's had always raced to convert a vision into a working device presented another bind. Weiser had come to PARC, he would later say, "hoping to integrate social science and computer science."[26] His conversations with Lucy Suchman were, by his own account, among those he found most generative. The demand for speed, however, seemed to offer little apparent latitude for collaboration between Weiser's lab and Suchman's anthropology team, who paced their ethnographical research around the slow burn of careful observation and thick description. The PARC anthropologists suggested that any efforts to design a new technology would likely yield superficial outcomes unless the technologists first endeavored to better understand the people and situations their inventions were intended for. To do so, the anthropologists believed, the timeline for major R&D projects should include a substantial period to study the workplaces into which the computer scientists hoped to cast their products.

Suchman's team was already one year into a meticulous ethnographic study at San Jose International Airport. Through countless site visits, interviews, and video recordings, they were examining how information circulated between air traffic controllers in the airport's central command room and all the staff on the ground who directed planes in and out of the gates. The project was funded by Steelcase, a leading office furniture company that had been eager to pay for fresh insights about emerging trends in the workplace. Suchman won them over with her pitch that airports were an ideal site to investigate the needs of employees in a fast-paced, highly distributed, and mobile work environment. The job of controlling air traffic was a dynamic, spontaneous affair, full of moving parts and tight schedules, loosely held in check by elaborate timetables, radars, and protocols augmented by ongoing adjustments relayed over two-way radios. From the staff seated in the command center to the runway crew referencing their clipboards holding carefully arranged documents, the airport workers wielded technology to filter information around the complex, high-stakes decisions they faced every minute. Weiser must have suspected there was much in the anthropologists' findings that promised to inform his computing agenda. But aside from some cross-lab chats, he noted that he "quickly vanished into [the Computer Science Lab]."[27] Thus, the airport ethnography and the effort to launch ubicomp progressed side by side without much of an exchange.

Weiser and Suchman were acting on different interpretations of the problem they had initially discussed. Digital interfaces remained highly insensitive to users and their surroundings; computer science modeled its work upon very loose approximations of people's actual lives. This insensitivity to context severely limited the capacity for so-called interactive technologies to understand a user's activity, as they often failed to register circumstantial variations unaccounted for by the machine's programmers. The final paragraph of Suchman's book had left readers with a question: "What are the consequences of that limitation?"[28] The main consequence, to Weiser's mind, was the growing gap between computing's virtual realities and the rest of the material world. He hoped to bridge that gap. The limitation could be overcome, he believed, through defiant and good-hearted feats of invention. Existing computer systems were insensitive to context—he would spearhead the development of new technology that would make them sensitive.

Suchman saw another solution, and pursuing it would alter the course of her research in the coming decade and aid the rise of a new field. It was present in the kernel of a sentence she wrote in response to her book's closing question, about the limits of interactive machines. "Practically, ingenious design combined with testing may do much to extend the range of useful machine behavior," Suchman insisted.[29] The breakthrough, as Suchman pictured it, would be not only technological but methodological. What was most needed was not revolutionary features, promising as they may be; a more impactful and socially beneficial revolution might result from changing how technical innovation happened. So much of the invention process occurred in labs sealed off from the very people whose lives technologists claimed they wished to improve. Suchman's suggestion to combine invention and design with ethnography—not just at the end of a gadget's development, but from the onset of its creation—ultimately implied that technologists should treat the living, breathing layperson as an equal partner in innovation (a notion that was already gaining traction in Scandinavia).[30] This was not a plea to merely *think* about users more. Good inventors always tried to do that. To truly involve prospective users in the design process, a technology firm had to first comb through the details of a user's day in the *user's* workplace, as Suchman's team was doing at the airport. Technologists themselves would need to become more attuned to particular contexts. Then and only then might they be in a position, Suchman reasoned, to design technologies that were sensitive to those contexts as well.

As sensible as that notion was, Weiser remained hesitant to sacrifice time in his lab—and its attendant creative freedoms—in order to engage with outside users and follow their feedback. Such practices sounded blasphemous to ardent disciples of the Computer Science Lab's traditions. Weiser's plan was simple, though not even he knew where exactly it would lead. His lab would stick with the usual methods that had worked so well for PARC during the '70s: *build what you want to use and use what you build*. According to this logic, a researcher's intuition held greater insight than any focus group remark or usability testing session. Weiser and his colleagues didn't entirely distrust user input; they just trusted themselves more. Inventors were users, too—and only by probing their own experiences could they feel their way through a nascent technology. To articulate a problem, they assumed, one needed first to encounter it with surprise and grow familiar with the distinct agitations it wrought. The precise nature of a novel sensation might get muddled by too much discussion early on. If engineers built the prototypes they were intrinsically motivated to build, then they would commit to learning from those exquisite tortures that plagued and delighted the creator working in solitude.

Their offices at PARC would be their canvas. Weiser's ambition to build "systems that touch the world"—systems that "fit with everyday human activity"—needed to be downsized, if only temporarily, in order to take shape as a manageable experiment.[31] Several factors made PARC an obvious and convenient test bed, homogeneous though it was. The project fit naturally with the research center's founding mandate to anticipate future workplaces and to invent products that might boost Xerox's future revenues. During a typical day at PARC, the computer scientists wore their office-worker hats about as often as their inventor hats. They participated in a workflow of documents, presentations, phone calls, and meetings that ostensibly resembled other offices elsewhere. Crucially, too, the place was packed with the best available technologies, not to mention all the homespun, in-house systems unknown to consumers. This set a high bar for novelty; it would be obvious if a proposed device too closely resembled something that already existed.

The novelty question raised some concern about "ubicomp"—at least, it did for the computer scientists who had signed on to the project. When they pressed Weiser on the nuts and bolts, he sometimes just pointed at gadgets that researchers had been developing at PARC or elsewhere. This was the starting point—"ubicomp phase I"—

he assured them. In the short term, Weiser contended, they could make real headway by bending existing technologies toward new uses. Companies across Silicon Valley had assigned engineers to take a stab at mobile devices, but that didn't necessarily mean PARC was playing the same game as them. Weiser told his colleagues, "The details of [devices] like this matter a great deal. . . . God is in the details, and only by building [the devices] ourselves can we get the details right."[32] While Apple, IBM, and the tiny computer startups competed with one another to market the best handheld PC, PARC would commit to developing and testing applications for handheld devices that no desktop could execute. Perhaps no viable products would result from this initial phase. Weiser dangled only the guarantee of a new mindset: "Using these things would then change us."[33] If little else, these things now had better names than "tiny," "medium," and "large."[34] Weiser now referred to them as "tabs, pads, and boards."[35]

Of the three, only boards got off the ground by the summer of 1990—and that was because the project had already commenced in another lab at PARC, under Scott Elrod's leadership. Boards represented the largest form factor in Weiser's ubicomp lineup, though Elrod's team continued to call them "Liveboards," the full name they had come up with first. The Liveboard loomed in PARC's common areas and conference rooms like a giant TV on wheels. It stood over six feet tall and four feet wide. Liveboard had spawned from the earlier Colab experiment and from recent talk around PARC about wall-size computing displays. The desire to optimize group meetings and improve collaboration ran high in many PARC scientists. PARC maintained a staggering human-to-whiteboard ratio. The break rooms had floor-to-ceiling whiteboards as walls, and every personal office came with a standard-issue whiteboard. When one colleague dropped by another's office for a quick chat, they gravitated to the whiteboard the moment they inched beyond small talk—they loved whiteboards and the spontaneous brainstorming that played out upon their surfaces.

The Liveboard team rallied around that feeling. A big interactive screen could show and do many things, but versatility was no longer a virtue in this case. Colab had been hampered by an overabundance of functionality; the more it enabled meeting participants to do, the further it strayed from its specific purpose. Liveboard removed Colab's semicircle of interconnected personal workstations from the room, as well as its dependence on a mouse and keyboard. Elrod and his colleagues noted, from years of daily experience, that "it [was] difficult to maintain the focus of a meeting when interaction with the central display is mediated by an adjacent keyboard."[36] While a keyboard and mouse were available if necessary, Liveboard users controlled its touch-sensitive display with a cordless electronic pen.

With these pens, multiple participants could each draw or write on the screen together. Three buttons on the pens could open pop-up menus for printing, saving, closing, or opening documents. A presenter (or any meeting participant) could also circle a word or flip to the next slide just by gesturing with the pen from several feet away. Compared to a mouse and keyboard, the pen afforded presenters a much greater latitude to move about the room and to privilege their interactions with meeting participants without losing their ability to control the screen. By incorporating only those features that seemed most pertinent to tasks associated with team meetings, the pen's design shed less-relevant functionality in pursuit of a more seamless integration between computing and context.

New software accompanied this new piece of hardware. The Liveboard was decidedly not a general-purpose machine like most desktop PCs. Its operating system was stripped down to four basic modes: meeting, slideshow, games, and bulletin board. Once you entered the meeting mode, the system cleared itself of anything that might threaten to distract from the live meeting you were having in the room. What remained was a blank canvas with easily selectable tools for drawing, writing, and annotating, with options to save, print, and share. Slideshow mode activated a separate toolbox that was geared toward presentations, and it cued the machine to be on the lookout for the presenter's swiping and scrolling gestures. Bulletin board mode transformed the screen into a communal platform for adding news items to be displayed in the common areas when no meetings were happening; in a similar vein, the games mode temporarily converted Liveboard into an arcade-like portal to pure play during a more social gathering (or whenever the researchers wanted a playtime break).

It was through the development of these activity-centered modes that the Liveboard really began to put ubicomp ideas into practice. The cordless pen had made computing power ready-to-hand for users; all they needed to do was pick it up and use it the way they used a dry-erase marker, as if the screen were a whiteboard. Liveboard's screen was interpersonal in its scale—it wasn't tailored to fit a single pair of eyes like a PC, and so no one had to peer over the shoulder of the person at the keyboard and wait for him to click on things. By virtue of Liveboard's streamlined software, the system offered very few reminders that it was in fact a complex digital machine. It was crafted to feel like a natural, inconspicuous extension of the office environment rather than a special piece of equipment that demanded an operator's undivided attention. A meeting conducted with the Liveboard, unlike Colab or MIT's Media Room, did not force any of its participants to assume the postures of a computer user.

Norbert Streitz, a visiting researcher from Germany who later galvanized the spread of ubicomp in Europe, was among the first outsiders to glimpse the Liveboard in the summer of 1990, when it was still kept in a secure room open only to Elrod's team, Weiser, and a handful of others at PARC. Reflecting on those encounters with Liveboard, Streitz called it "the furthest deviation from a standard PC at that time. . . . Standing in front of this large, vertical display and reaching out with your arms in order to interact with a pen was quite a new experience."[37] In Weiser's view, Liveboards—boards—were just the first rendered image of an elaborate collage he was sketching in between meetings with Elrod's team and the teams who were building pads and tabs, the other two ubicomp prototypes.

Pads and tabs entailed some technical challenges that boards did not. Foremost, CSL researchers would be hard pressed to make a notepad-size computer (pad) or a postcard-size computer (tab) do any-

A prototype of a PARC tab (*left*) and a diffuse infrared transceiver (*right*) used for wireless networking at PARC. Photographs by Roy Want.

thing interesting without draining its battery in minutes. Weiser took a page from computer history and urged each team to fake that part for the time being. PARC's pads and tabs would operate initially as mere terminals, just as time-sharing monitors had done before PCs took off. Eventually, Weiser believed, Moore's law—which postulated that the number of transistors per silicon chip would double every year while the cost dropped—would allow PARC to cram all the necessary components into a small device. For now, pads and tabs would seem autonomous in use; but in reality, their innards' sole achievement would be to maintain a wireless connection with larger machines nearby. Software created for pads and tabs was hosted remotely on file servers and desktop workstations throughout PARC's facility.

In addition, neither PARC nor anyplace else at the time had Wi-Fi or Bluetooth technology—the standards that enabled these forms of short-range wireless communication had not yet been developed. To connect tabs and pads wirelessly to each other, the lab hacked together a local network out of near-field radio and infrared transceivers that would've seemed a lot closer to Wi-Fi than anything else on offer during the early 1990s. In order to get an application stored on a server up and running on a pad or tab, PARC servers passed data over the Ethernet to desktop workstations, which were in turn wired to a ceiling-mounted transceiver. Each transceiver—there was usually one per room—

emitted a wireless signal that could then be picked up by pads and tabs in the same room. As more rooms around PARC eventually had radio and infrared transceivers installed, the scientists could carry pads and tabs with them almost everywhere they roamed throughout the facility, using them without lag or interruption, since another transceiver was typically near enough to preserve their wireless connection with the building's Ethernet. Configuring the infrastructure to support pads and tabs involved a maze of technical undertakings. Once completed, however, it turned PARC into a kind of time machine. The computer scientists could (so long as they stayed in the building) make mobile devices do things they could not do elsewhere.

Though Weiser had no intention of bringing pads and tabs to the consumer market anytime soon, the applications he and his colleagues were creating for them at PARC made seemingly similar devices—such as the Poqet PC, as well as the handheld that Apple was then working on—look like glorified calculators. Apple's 2007 iPhone would in retrospect bear little in common with the Apple Newton, the company's allegedly state-of-the-art tiny computer, which began in their lab during the late '80s before flopping with customers in 1993. By contrast, a 1991 promotional film showcasing Weiser's group and their ubicomp prototypes went on to astonish tech enthusiasts when the footage got uploaded to YouTube in 2009.[38] The film, scripted by Weiser, would lead these twenty-first-century viewers on a guided tour of PARC inventions that appeared to resemble their current smartphones and even portrayed applications yet to be implemented by the day's leading tech companies.

Marred by its low production value and aching for better actors, the 1991 promotional film still managed to show how tabs, pads, and boards might alter office workers' relationship to digital information as well as their physical environment. "Potentially numbering in the hundreds per person," the film's voiceover begins, "these devices are nothing like what you use today—they are mobile, they know their location, and they communicate with their environment."[39] Most people hearing this in 1991 would have been too flabbergasted by the first part of

Weiser's statement to genuinely process the rest of it. Saying there'd be hundreds of computers per person sounded about as outlandish as Licklider's notion, from back in the '50s, that everyone would someday want their own computer. In each case, though, the absurdity faded when you realized that they (first Licklider and now Weiser) had in mind a different definition than their contemporaries did. Licklider's personal computer wasn't the hulking mainframe apparatus that his peers assumed to be the pinnacle of electronics. It wouldn't occupy half of someone's house, just their desk. And attending to Weiser's ubicomp ensemble—the dispersed array of tabs, pads, and boards under each user's purview—wouldn't be like using a hundred PCs. To understand why, you had to understand the significance of Weiser's last phrase: ubiquitous computers *communicate with their environment*. You had to understand the tab.

Tabs were the smallest yet most sophisticated of the three ubicomp building blocks. Of the hundred or so computers inhabiting a typical room in Weiser's vision, at least 90 percent would be of this inch-size variety. Tabs—groups of them and never just one—held the key to context sensitivity. Through them, computing could inhabit the nooks and crannies of everyday spaces; their software could register what was happening in a room and adapt its operations accordingly. Different as the Liveboard was from a desktop workstation, it was still a stationary machine that fit only in so many places. Tabs continually reshaped their content in direct response to live actions taking place in the vicinity. Tabs moved with people. The scientists put tabs into their pockets, clipped them onto their jean waistbands, and carried them in briefcases without a second thought. Tabs also attached easily to other objects. Inside PARC, tabs were attached to the side of coffee machines, hung on office doors next to researchers' name plaques, or were left forgotten in the crevices of couches. They spread around the building like points on a shape-shifting constellation, each day's end revealing a fresh portrait of the lab's people and things.

Connectivity became the baseline condition around the office; digital traces of the computer scientists' plans and whereabouts now adorned the building's interior. A new example awaited around every corner. The tab's calendar app could sync with the calendar on your desktop, the way a Poqet PC could, but the tab's calendar would also share information with other tabs passing through the vicinity. Simply walking by a conference room, for instance, cued the walker's tab to display the meetings scheduled for that day, week, or month, along with a link for booking the room. Passing by a colleague's closed door might show you the parts of his personal calendar he had made public. Another tab app in the works aimed to give PARC visitors a guided tour of the campus, offering up text or audio descriptions about the research happening in each wing of the facility as visitors walked into that space. Apps of this sort meant to demonstrate how the tabs, in concert with pads and boards, might plug their beholders into a tacit ecology of otherwise invisible relationships between objects and people, individuals and the organization, inner lives and shared spaces. Tabs charted a course upon which computing might reinforce one's sense of connection and community; they offered a means through which the interpersonal ephemera conjured in Heidegger's notion of "entanglement" could be traced, recorded, and digitally presented. The "sea of bits" a user accessed alone on her PC was, via tabs, made to flow into a series of canals whose surfaces held reflections of the land at its banks. "Context," the PARC Tab team wrote, "can be used to filter information. Instead of presenting the complete file system . . . [the screen] shows only files whose information is relevant to the particular room it is in."[40]

Groups of ubicomp devices could also filter information not just according to place but according to time. Among other things, this created a more conscientious way to manage emails. The researchers could set their tabs and pads to notify them of incoming messages only at times when they appeared to be by themselves—that is, when no other person's devices were present in the room. Any type of data (in this case, emails) could be regulated by the onset of certain states of being, which each scientist could program as she saw fit. The imaginable possibilities were many and amenable to all sorts of eccentricities:

When I enter a restroom, pull up this evening's local weather forecast; *when I enter conference room A, allow my colleagues to see that I'm busy with a meeting in conference room A*; or *when I am sitting with two or more people, change my display font from Times New Roman to Helvetica*. Another experimental app called Forget-me-not "provide[d] a tab user with an automatic biography of their life."[41] If you so desired, Forget-me-not would generate a day-by-day chronological list of everywhere a user went in the office, who she met with, what files she printed, and all her phone calls. The app served as a memory aid, and each user could select from preset icons to search among the recorded events. Furthermore, if a room was outfitted with tabs, pads, and boards, then the scientists could control aspects of the room by tapping on their handheld screens. Meeting participants could gather on beanbag chairs around a Liveboard and write, draw, or click on it using the pads in their laps. If there were five people with pads, then the Liveboard would display their five respective cursors, each one moving around on the big screen as each person moved a finger over their smaller screen. Such capabilities were all the more expansive on a tab-to-tab level.

With a critical mass of other tabs installed nearby, any given tab could become a "universal controller."[42] An app called Remote Control essentially let authorized users alter the environment the way one changed channels on a TV—just press a few tab buttons to adjust a room's lighting, temperature, or presentation setup. It was becoming increasingly clear that tabs, pads, and boards were not only new vehicles bringing computation into the physical environment. They also held the promise to redraw entrenched boundaries that had long stipulated the rules of engagement between human will and the built environment. It seemed, at this rate, that clicking on a screen might someday execute instant material changes to settings of all kinds on a scale unapproachable by other means.

Weiser's excitement about the tab's prospects grew with the arrival of two new hires he and Brown had recruited toward the end of 1990. Until then, many researchers in the Computer Science Lab had been drawn by the lab's unparalleled track record in personal computing, and they were handpicked for their strengths in that area. Technical expertise could be easily redirected in the service of new goals, of course; but committed scientists didn't relinquish their deepest passions at a lab manager's request. The two new recruits, Rich Gold and Roy Want, were the first to join because of Weiser's vision.

In every other way, Gold and Want were about as different as two short white men could be from each other. Possessing a collected demeanor and impeccable credentials, Want gave the impression of someone who had grown up strolling from one science experiment to the next in khaki slacks and neatly tucked shirts. A Londoner who acquired a bachelor's and a doctorate from the University of Cambridge in less than eight years, he was kindly articulate and soft-spoken, though his work generally made it clear that he was the smartest person in the room. Want was by his late twenties one of the world's foremost authorities on mobile hardware design.

Gold's early attempts to dress the part left him looking like a man who had wandered into PARC wearing clothes stolen from a Stanford dorm room—he walked the halls in striped rugby polos, black jeans, and tennis shoes. He came to Silicon Valley by way of Buffalo, New York, and art school. Were it not for his compulsion to push the boundaries of music and literature, he never would've fiddled with the synthesizers and software that led him into computing. His assorted background as a composer of electronic music, performance artist, writer of handwritten manifestos and progressive children's books left unpublished had gotten him in the door at the toy company Mattel in his late thirties, where he catapulted up the chain to become senior design manager for the company's new computerized toy division "seeking new and unusual uses for small computer systems."[43] By the time he arrived at PARC, Gold had mastered the art of charmingly selling himself and the far-off galaxy of influences—kinetic sculpture, dadaism, automatic music, vintage lunch boxes, Guy Debord—that he smuggled into R&D meetings.

Want and Gold were, in sum, exactly the collaborators Weiser needed and wanted. Their tandem presence supercharged the lab's technical capacity and breathed new life into the ubicomp imagination. By spring, Weiser felt like their progress was outgrowing the building. The thought of what they might create together underscored the limited scope of their office-bound concepts. The initial projects had been interesting and often exhilarating, but these just tickled the surface of Weiser's ultimate dream. Actualizing pieces of it left him itching for more: more resources, more funding, more prototyping, more conceptual breakthroughs, more news coverage, and even more fresh talent. The project seemed on the cusp of an exponential moment. It wouldn't last long enough for Weiser to tell the world about ubicomp one MIT Club at a time. PARC needed a sweeping gesture to plant its flag in this future—now, while the battle of ideas about tiny computers was still undecided—before the tech industry hitched itself to another vision. On a dewy morning in March, Brown greeted Weiser with an opportunity.

Weiser brought home a special assignment in the spring of '91, and with it he had turned his family's Palo Alto home into a second R&D facility of sorts. "The house was completely overrun with tabs and pads and orange extension cords," recalled Weiser's wife, Vicky Reich.[44] There were tabs stuck to the refrigerator, pads strewn about the living room. The house of course lacked his lab's novel wireless infrastructure, so the devices in the kitchen had to be connected by cables to those located in the bedrooms, the garage, the dining room, and so on. "It was fun," Reich said. "The house became a lab for what ubiquitous computing might be in the home."[45] The whole family joined in. Weiser looked on as his youngest daughter, Corinne, tested CSL's stylus-based handwriting software as she explored apps on the tab.[46] She and Nicole, Weiser's older daughter, had begun spending summer days with their dad at PARC, where he and his collaborators loved to watch them try out new technologies and listen to their feedback.[47] At home, the prototypes lost a lot of their functionality when removed from PARC's intricate network. But that was fine. Weiser's task around the house was different from one he was pursuing with colleagues at the office. Carefully arranged messes of books, pens, and sticky notes took root wherever Weiser sat and thought.

At home, Weiser threw himself into writing an article that could be, he suspected, the most important project he had ever taken on. He kept his notebook on him throughout the evenings and weekends so that he might jot down ideas as they arose. Having the pads and tabs everywhere in plain view lent an air of reality to the questions he was wrestling with: What would ubicomp feel like once it leapt out of his lab and into the world? What kinds of things would people like tabs, pads, and boards to connect them with throughout their day? How could he get a lay audience excited about the idea of being surrounded by hundreds of connected devices in every room? The blank page before him introduced concerns, doubts, and second guesses that had not burdened the rush of his first thoughts.

The audience Weiser was writing for dwarfed any he addressed previously in his industry presentations and academic essays. The argument he was now trying to outline would appear, pending final approval, in the exalted pages of *Scientific American*, his father's long-time favorite. On the floors of Weiser's teenage Stony Brook home, back issues of the magazine had amassed in stacks that looked like little paper towers. The current editor had approached Brown about contributing to a special issue on the future of computing, and Brown in turn approached Weiser. It was their chance to gauge how big ubiquitous computing might become; whatever expression Weiser's words managed to give his vision would be sized up against the best. Some of the world's leading researchers were already set to be involved, including Alan Kay and Nicholas Negroponte, alongside then vice-presidential candidate Al Gore. This special issue, set for publication for September 1991, was bound to interest a giant swath of readers across the nation who were curious about where technology might be headed. Silicon Valley's exploits were reaching new heights in the country's popular imagination. While only a tenth of American households owned a computer, major media outlets brimmed with speculation that some grand transformation was on the horizon.

The future now stirred in Weiser a new mix of feelings. It was no longer a slow, organic melody unfolding at the pace of his own progress. Competing visions were pressing in from Silicon Valley and elsewhere. The machines of the day—against which he defined the merits of PARC's tabs, pads, and boards—were starting to undergo promising changes themselves. Computer scientists more accomplished than Weiser had big plans for the familiar screens that he was hoping to coax people away from. Just as Weiser was plotting ideas in his living room, Silicon Graphics founder Jim Clark was in a hospital bed a few miles away, recovering from a motorcycle accident and finishing up a conference paper he titled "A Telecomputer." Clark's previous innovations in computer graphics had set the standard for high-end workstation displays in the 1980s, and he now believed the '90s would be defined by the marriage of television and computation.[48] Clark's envisioned consumers would simply upgrade their TVs and gain sudden access to everything that networked computing stood to offer them: "movies on demand, virtual reality games, digital forms of daily newspapers, monthly and weekly magazines, libraries, encyclopedias and interactive books."[49] The TV would thus become a supercharged PC in disguise.

Around the same time, in the Swiss suburbs of Geneva, researchers working at CERN—a massive laboratory renowned for the physics experiments conducted with its state-of-the-art particle accelerators—were experimenting with an online information repository they were calling "WorldWideWeb." CERN, as a hub of collaborations involving physicists around the world, had since 1984 supported in-house R&D efforts to create protocols that would enable scientists to share their data more readily on a global basis. Whereas the laboratory's early networks had been approved to link only its on-site computers, the WorldWideWeb project was the first CERN network that benefited from the organization's policy change, in 1989, to allow for external connections over the internet. The WorldWideWeb was built on a single NeXT computer by CERN's Tim Berners-Lee, and its first users could access it only from NeXT machines. Initially, there was nothing worldwide about it. The project's main page, a few chunks of plain text on a white background, stated that Berners-Lee's ultimate objective was "to give universal access to a large universe of documents."[50] Nearing its fifth month of existence, the WorldWideWeb was becoming known to dozens of scientists at CERN and a smattering of universities that shared files over a handful of online pages.

A more popular platform, LambdaMOO, had just launched down the hallway from Weiser's office. His friend and PARC colleague Pavel Curtis had created a "text-based virtual reality" modeled on the layout of his Palo Alto abode.[51] Like Weiser, Curtis logged many hours on Usenet's online messaging boards; he heard in them a cry for something more than information swapping. Online networks were capable of far more than that, as a handful of virtual community-building experiments had recently shown. LambdaMOO wasn't the first such effort, but it quickly boasted heavy traffic. On LambdaMOO's start page, Curtis billed his new cyber dwelling as a "new kind of society" in which users typed directional commands to navigate among various "rooms"—the Coat Closet, the Living Room, and so on.[52] Entering a new room opened a paragraph that described the room: "It is very bright, open, and airy here, with large plate-glass windows looking southward over the pool to the gardens beyond."[53] Each space doubled as a chat room where assembled users typed out messages back and forth in real time under the cloak of usernames like Juniper He, Zippie, and Mr. Bungle. LambdaMOO users came to socialize and partake in the building of an alternative world that allowed for "modes of behavior not usually seen 'IRL' (in real life)."[54]

These new directions in tech seemed like a set of possible chess moves being considered in the industry's game of winner-take-all innovation. Each move opened a set of novel possibilities, all while diminishing the prospects of other paths that would remain dependent on moves not taken. The possibilities lost in a slew of shortsighted moves might never come about again. Almost every technologist Weiser knew outside PARC remained committed to ushering more of life onto the screen, as if it were a foregone conclusion that, like them, untold masses of people would soon be staring at PCs for most of their waking hours.

Computer scientists of all stripes, in academia and at corporations, still wished to expand the scope of each desktop by connecting users to a plethora of shared databases. The contents of one PC were being made available to every PC. It had begun with electronic documents, with the linking of many texts across many computers, but gradually the hardware would evolve to reliably handle audiovisual media and, perhaps much later, the full-fledged virtual reality simulations then being tested by researchers on clunky headsets in a few lavishly resourced university labs. At some point along this trajectory, these nascent galaxies of online resources might grow too big to ignore and too essential to displace. Ubicomp would be a much tougher sell then. Weiser had to start making his case against all the technologies that led toward virtual reality before they became the norm.

Weiser found himself jotting down sentences for his article that mourned the loss of being with others in a shared physical space. He lamented, "Even today, people holed up in windowless offices before glowing computer screens may not see their fellows for the better part of each day. And in virtual reality, the outside world and all its inhabitants effectively cease to exist."[55] Yet, the pages in his notebook surrounding these passages revealed another more personal, ironic conflict playing out in Weiser's head as he was formulating his argument.

Sandwiched between sentences differentiating ubicomp from virtual reality were comments he had written during a recent off-site retreat with his fellow PARC lab managers. One page recorded a list of complaints about him, which had been submitted anonymously by a few computer scientists who worked in his lab. They accused him of roughly the same criticisms he was leveling at virtual reality:

> "Mark, you have a talent for ignoring people's input."
>
> "Mark, you ignore people's concerns with optimistic priorities."
>
> "Mark, you promise everything and deliver nothing."[56]

These retreats had become a regular facet of every lab manager's job. As PARC director, Brown hired executive coaches to conduct workshops meant to help managers understand how their personal psychology shaped their leadership style. Facing up to peer feedback was a key exercise to guide one's development. Pam Roderick, a former social worker from England who was the staff's favorite coach, retained a distinct memory of Weiser from their first workshop together. "It was hilarious," she said, "Mark had his feet up on the table. I was giving the managers some pretty awful feedback about what it was like to work for them. Then Mark looked at me and said, 'Well, that's obviously their problem!' And I looked at him and said, 'If you do not get that this is your problem, then we are wasting our time.' He pulled his feet off the table and sat up, and that was the start of it. . . . Mark was one of the ones who came around most quickly."[57]

After the retreat, in the privacy of his notebook, Weiser attempted to make sense of the feedback he had received—namely, that some colleagues felt ignored by him. The apparent cockiness with which he greeted the news during the workshop swiftly collapsed. He listed a few of his underlying traits that seemed like relevant scapegoats on which to pin the criticisms: "Me: Independent. Self-Reliant. Under no one's control."[58] There was some truth to this assessment. In between his many roles at work and home, and in spite of the ways his writing bemoaned electronically induced solitude, Weiser insisted on being alone regularly for long stretches. "He was an introvert," said Reich. "He really needed quiet time to regroup. He really, really desperately did."[59] Still, while taking his breaks from people, Weiser's notes to himself so often reiterated a longing for greater interpersonal connection, as they had back at New College. Just below the list asserting his autonomous nature, Weiser wrote a subsequent list, outlining other habits of his that clashed with traits in the first list:

> One: speaking flamboyantly encouragingly to the lab, explained by my need for them to hear me (might be ignored).

> Two: reluctance to press forward hard, urgently, with my ideas (explained by my being at my followers' mercy).
>
> Three: fear of being ignored, forgotten (left at home for family trip).[60]

To clarify the depth of that last fear for himself, he added: "can't believe anyone would rescue me from fire."[61]

The margins of lab notebooks he kept throughout his time at PARC regularly bore agonizing traces of the panic Weiser evidently felt upon sensing any fissure in his relationships. He fretted about possible miscommunications, wondering if he had said something wrong, interrogating the meaning of what others said to him as he struggled to piece together the remembered bits of conversation he scribbled down. Those were the moments of a meeting that tended to stick with him, even when they were just blips in an otherwise harmonious exchange. Much more often than not, his colleagues would use superlatives when they described the warmth, sincerity, and zeal many felt in his company (though the last of these qualities could morph into "stubbornness" if your goals competed with his). Many would tell his daughter Nicole about how he had impacted them, how he had helped or comforted them in some important way.[62] Scientists around PARC who were unaffiliated with Weiser's lab would occasionally flock to his corner of the building just for an injection of good vibes. "He was very compassionate and one of the most forthcoming people at PARC," said Scott Elrod.[63] It never would have occurred to most of his colleagues that he could harbor the depths of self-doubt reflected in his notebook entries. Wherever Weiser went, from Stony Brook to New College to Palo Alto, a pattern appeared to follow: people often felt more connected to him than he seemed to realize. He could be quite reticent to trust the strength of his bonds, even with those who clearly held his friendship dear.

What, then, were these connected tabs, pads, and boards he saw cluttered around him? At PARC, these were the building blocks of a new paradigm that might reshape the flow of digital information. They appeared differently now that they were sitting around his house, revealing perhaps another, deeper dimension. Wires from tabs and pads crisscrossed the rooms like ropes wrangling everyone loosely together. They ran in and out of doorways, stretching back and forth to reach some constant co-present moment.

For months, Weiser labored over small sections of his article as he shared rough drafts with peers at PARC. The revisions carried well into summer, spurred on by his conversations with Brown and the philosophy books of his youth that he now returned to with greater purpose. A realization must've sunk in by the time he proofread the piece one last time, surrounded by prototypes. The momentum that had been gathering around his vision would either boom or stall from the reactions to this article. Success on that level would be eminently more noticeable; but so would be any ridicule that might accompany a public failure. In any case, *Scientific American* readers would be shown photos of the neatly crafted illusions of ubicomp on display at PARC. His writing gave no hint of its current state in his living room.

5

One Hundred Computers per Room

"THE BEST ARTICLE (ATTACHED) is the one by Mark Weiser of Xerox PARC, which I think everyone should read," wrote Bill Gates in a memo to his Microsoft executives. "The rest of the articles are interesting but not critical."[1] The September issue of *Scientific American* was still being delivered to mailboxes and newspapers when Gates typed these words on August 28, 1991. His memo had quickly been copied by someone at Microsoft and printed by someone at Xerox PARC, where it landed on Weiser's desk. As Weiser read the software tycoon's flattering decree, he must've let himself marvel at how big this vision of his might actually be. A month full of doubts evaporated in an instant.

Weiser's article had nearly been killed in the final steps leading up to publication. The magazine's editors were thrilled by his ideas, but they were disappointed to learn that he was the sole author. They had invited John Seely Brown. Brown had name recognition. As PARC's director and a fixture on the industry's speaking circuit, Brown was almost as famous as the likes of Nicholas Negroponte, Alan Kay, and the other star contributors the editors had assembled. *Scientific American* feared that a lesser-known name like Weiser's might make their special issue a little less special. They responded with an ultimatum: they would publish the article only if Brown's name was added as a coauthor.[2] Weiser and Brown had indeed passed a few pages back and forth, as it was initially their plan to collaborate. While the piece grew from their conversations, the final outcome was, Brown insisted, "so much more Mark."[3] Brown got on the phone, and the editors gradually relented, though they still worried that printing only Weiser's name would lead readers to skip over the article.

Gates had not skipped over it, and, thanks to that memo now in Weiser's hands, no one at Microsoft could safely ignore it, either. That fall, Weiser's mailbox filled with a mess of postcards from computer scientists around the world asking Weiser to have Xerox mail them a reprint of "The Computer for the 21st Century."[4] (Weiser's colleagues at PARC sensibly nixed the obtuse title—"Embodied Virtuality"—that he first proposed.[5]) The postcards bore addresses from every region of the country and many cities across Europe and Asia. Some researchers had come across Weiser's article in the library and wanted a copy to study at length. Others referenced the word-of-mouth buzz that was spreading around their companies and campuses. Then a larger, second wave of requests poured in after the *New York Times* ran a hype-heavy piece about ubiquitous computing, one intimating that PARC had again divined technology's future. Weiser's unremarkable epiphany in the shower had now matured into a headline-grabbing R&D agenda that, according to the *Times*, "would help to keep the United States competitive in high technology during the 1990s."[6] Journalists from the *Washington Post* and the *Chicago Tribune* followed with stories of their own, undeterred by Weiser's estimation that ubiquitous computing would be a twenty-year project.[7] They itched to break news about any next big thing and they gave the impression that tabs, pads, and boards were coming soon to stores everywhere.

The lines from Weiser's article that captivated readers and reporters were the same ones that made Silicon Valley strangers crowd around his table at the Dutch Goose, a dive bar near Stanford where Weiser held office hours for the curious in light of his dawning fame. The Dutch Goose was a place where intellectuals came to get rowdy. Students witnessed a side of their professors not shown on campus. Scientists cussed as they shared laughs about the mysteries of the universe. The long wooden tables were covered in names, initials, equations, quips, and drawings that customers had etched into them. Over deviled eggs, pitchers of beer, and sandwiches dripping with grease, Weiser expounded on his published words as new fans gathered around to ask about the ideas in his article. Friends of friends kept coming and going, like a revolving door of would-be disciples.

They all, like the press, were smitten with the article's opening sentence, even if they weren't sure what it entailed. "The most profound technologies," Weiser had written, "are those that disappear."[8] To illustrate the point, he had readers consider the evolution of writing and its culmination in the small miracle of candy wrappers. The packaging that enveloped Juicy Fruit gum and Milky Way bars were not only, in Weiser's eyes, watershed artifacts in the history of civilization. He ventured that they could teach the tech industry a lesson that might rescue computers from their current state.

Previously, if technologists had gleaned any adages from writing's long history, it was that having more information was always better than having less. Weiser lacked room to elaborate on this point in writing, but he probably raised it more than once during barroom digressions. A founding assumption had stretched unquestioned—from Vannevar Bush's postwar memex machine to the hypertext programs of Tim Berners-Lee's WorldWideWeb experiment—that electronic screens ought to double as vast libraries. The dream surfaced first in a short essay by the sci-fi novelist H. G. Wells called "The Idea of a Permanent World Encyclopedia." In 1937, Wells marveled at the possibilities raised by a new device in American libraries called microfilm. Microfilm shrank books and articles to a fraction of their printed size; massive collections of texts reproduced in this manner could be sent quickly and cheaply around the globe. For the first time in human history, it occurred to Wells, we could bundle all recorded knowledge into a massive encyclopedia and distribute it to any library that owned a microfilm reader. Wells reckoned that microfilm might eventually serve as a kind of artificial brain that would "pull the mind of the world together."[9] Stepping foot inside your local library and sitting down at such a machine would thus connect you to every single library all at once.

Candy wrappers—the form of writing Weiser's article praised (along with street signs and graffiti)—were about as far as one could get from libraries without ditching the written word. By elevating the candy wrapper above the library as a source of technical inspiration, Weiser was attempting to cull from history to suggest an alternative foundation for digital innovation. Candy wrappers and street signs were profound, he argued, because they made words disappear into—*become a part of*—the objects and environments they described. (The word *disappearance* in this sense, Weiser noted in a quick aside, derived from his readings of Polanyi and Heidegger—that is, he meant to signify what Polanyi meant when he said that a white cane user feels the street, not the cane; the cane "disappears" as it connects one's hand to a series of sensations stemming from the ground.) For all their riches, libraries did the opposite: they were buildings constructed to hold bound volumes of words in a space intentionally sealed off from other settings.

Weiser had nothing against libraries, of course. He was married to a librarian and loved surrounding himself with books. It was his habit to jump between many books on assorted subjects all in one sitting—ever since high school, during any given week, he could be found making his way through upward of a dozen volumes, each within arm's length.[10] But to mistake books, magazines, and newspapers for the only worthwhile achievements of written culture, Weiser contended, led one to neglect the absolutely crucial roles played by other seemingly petty genres. The dominance of the library metaphor kept technologists blind to a wider spectrum of literacy innovations that might inform their designs. Once you appreciated "the real power of literacy" that candy wrappers embodied, then you could appreciate the need to develop digital interfaces that were modeled after their merits.[11]

Industrialized cities and towns, Weiser's article had pointed out, came to be smothered in useful words throughout the twentieth century with "street signs, billboards, shop signs, and even graffiti. . . . Candy wrappers are covered in writing."[12] These texts shared a unique attribute: they communicated their messages to readers on location, at precisely the moment readers were likely to want that information. To really grasp the difference between a book and a candy wrapper, all you needed to do was imagine a grocery store in which every product on the shelves was devoid of writing—no price tags, no branding, no nutritional labels, no aisle markers. Imagine that printed words and numbers appeared only in books and nowhere else. How would shoppers go about their shopping? If the grocery store was set in a literate society that published information only in bound volumes, then the only logical thing to do (as absurd as it would seem) would be to give every shopper a hefty, phonebook-like directory containing information about all the products on the shelves. Want to know the price of that gallon of milk in front of you? Simply turn to the "M" section of your directory and find the image of a milk carton on the page that most closely resembles the shape of the carton in question. Curious about the sugar content of that jelly jar? Just flip to "J"!

The frustrations and limitations of a hypothetical books-only world would mount tenfold in settings like airports, bus stations, and highways—places where society had grown accustomed to consulting gate numbers, destination listings, and road signs posted where we need them. Fortunately, road signage such as exit markers distributed navigational information on a geographical basis so that it was situated in plain view along the highway. You didn't need to keep a book in your lap while driving to ascertain this basic information. Weiser noted with admiration that road signs, candy wrappers, and the like "[did] not require active attention."[13] Emblematic of technologies at their most profound, words written directly upon the objects and locations they described effectively "disappear"—they "[wove] themselves into the fabric of everyday life until they [were] indistinguishable from it."[14] Personal computers, by contrast, were conspicuous and unrelenting. Weiser segued from the glories of candy wrappers to the shortcomings of PCs. The irksome thought experiment of removing all such texts from modern society suggested a far more irksome realization: computers were pulling us to regress in that direction.

Sitting at a desktop workstation had much in common with the scenarios one could imagine if print culture had never evolved anything other than the book. Just about every computer aside from the context-aware tabs and pads being tested inside PARC was a placeless repository for software and files that bore no meaningful relation to any physical setting, let alone its user's present surroundings. Electronic information could flow from one PC to another as if from one book to another, but it did not mesh with the environment the way printed price tags attached to items in a store did—or the way PARC's Liveboard project aspired to tailor its screen to fit the dynamics of a group meeting. Mainstream digital culture lagged well behind writing in this important regard. "More than 50 million personal computers have been sold, and the computer nonetheless remains largely in a world of its own," Weiser insisted.[15]

Neither his *Scientific American* readers nor the bar crowd at the Dutch Goose would've been hard pressed to disagree; indeed, it may have never occurred to them why this should even be considered a problem. The early promise of cyberspace was the promise of a new frontier—a welcoming, alluring break from established social norms. But when you pondered the issue with Weiser's analogy in mind, an uneasy paradox grew more certain: partaking in a global village of electronic connectivity would temporarily disconnect you from everything else. It wasn't so much of an issue at present. However, if computation continued to hold firm to the PC model, then the gap between our screens and the rest of the world stood to widen immensely as more people spent more of their day online. Virtual realities would scarcely reflect the actualities of life off screen, and you would be caught in the middle trying incessantly to toggle between.

After conveying a sense of this impending divide, Weiser made it clear how incompatible he believed current trends were with his ultimate vision. He declared in the article that "the idea of a 'personal' computer [was] itself misplaced"—that desktops, laptops, and handheld workstations together represented "only a transitional step toward achieving the real potential of information technology."[16] That potential would never be realized as long as computers remained "a demanding focus of attention" and kept people "holed up in windowless offices."[17] If the digital revolution had any hope of enhancing worldly engagement and corporeal interaction, that hope lay in "drawing computers out of their electronic shells," wrote Weiser.[18]

Tabs, pads, and boards constituted a first step in this direction. They were not yet capable of disappearing into the environment to the extent that Weiser wished, but they did a better job of it than PCs, and this hinted at what might become possible next. While Weiser suspected tabs, pads, and boards were not the ultimate vehicles of his vision—calling them "phase I" around his lab—he wrote about them as if they were essential for the time being. Weiser told his readers that *hundreds* of tabs, pads, and boards would occupy a typical twenty-first-century room. He enumerated the many uses to which PARC scientists were already putting these new devices around their lab, and how, moreover, they could be attached to a belt buckle or displayed on a wall as discreetly as a sticky note, a clock, or any other of the hundred-plus textual media that office workers so naturally relied on.

Once Weiser felt that his references to writing had sufficiently eased people into the idea of having computers everywhere, he turned the analogy upside down. Familiar comparisons gave way to radical contrasts. Tabs, pads, and boards, though "just the beginning of ubiquitous computing," could already do a lot more than attach words to things.[19] Unlike candy wrappers, they were dynamic and able to change their state, gather fresh data, and show new content in response to live action happening in the vicinity. Tabs, pads, and boards seemed to answer Socrates's ancient complaint about writing—that a piece of text remained silent when you asked it a question. While Weiser knew these devices were a humble start, he had an eye on blurring the boundaries that separated the living from the lifeless: "What will be most pleasant and effective is that tabs can animate objects previously inert."[20] To offer a sense of how these newly animate objects might improve life in the future, Weiser summoned up whatever narrative skills he had cultivated during his brush with creative writing back in college. He ended his article with something of a short story that chronicled a morning in the life of a fictional, twenty-first-century Silicon Valley technologist he called "Sal."

From her bedroom to the kitchen, from her garage to the office, Sal moves through spaces in which "almost every object either contains a computer or can have a tab attached to it."[21] The way a candy wrapper attaches words to chocolate, tabs attach digital media to everyday things in Sal's home. Sal's every action triggers a reaction she welcomes from the discreet interfaces in her midst. Data accompanies her almost everywhere she goes, and she would feel alarmed if a moment's glitch led the network to lag behind. She expects to see wherever she looks information that is relevant and in sync with the physical scene playing out before her. She has come to depend on hundreds of computers tucked neatly into the folds of her daily routine. None are gadgets she must fiddle with; none contain artificially intelligent software programmed to make decisions on her behalf. They are to Sal more like a sixth sense, existing as a finely engineered aid to her intuition. Whenever a perception gives rise to thought and a thought falls into speculation, Sal glances at electronic displays never far from her gaze. Digital captions offer additional means by which she can better understand her surroundings—the place and the people, what has happened recently, what might happen next. Objects around her house and the building where she works endeavor, in turn, to attune themselves to her.

It begins before she is even conscious. Her alarm clock notices her rolling around in bed and correctly interprets the movement as a sign that she is about to wake up. The clock gently asks the half-asleep Sal, in the manner of a knowing parent or spouse, if she'd like some coffee. When Sal mumbles yes, her clock tells her coffee maker to start brewing.[22] Coffee in hand (the first of many cups Weiser will have her drink this morning), Sal lingers at her bedroom windows.

Windows are where the most advanced technology resides in her world. Part translucent glass and part computer screen, the windows facing the street frame Sal's view of her sunlit neighborhood. The windows also display "electronic tracks" that show traces of where her neighbors have been out walking earlier that morning. "Time markers and electronic tracks on the neighborhood map," Weiser added, "let Sal feel cozy in her street."[23] Another window facing the bedrooms of her children displays a record of their movements, too. Their electronic tracks indicate to Sal that they are awake and now in the kitchen. The kids, from their vantage point, can also see visualized data attesting to Sal's whereabouts. The family's keeping tabs on one another is portrayed as a gesture of intimacy: "Noticing that [Sal] is up, [her kids] start making more noise."[24] The graphical trails on the windows are, in Weiser's telling, like extended facial expressions that clue people into each other's condition from a distance. The kids no longer need to guess at whether they might be disturbing Sal's sleep; the windows give them a general sense of what she's doing behind the closed door. These electronic tracks, in Weiser's depiction, constitute a medium of family communication—surveillance in the service of love. The kids want to know what the graphics tell them in order to know Sal better, to feel closer to her.

Sal makes her way to the breakfast table and reads the newspaper. A sentence from a column in the business section piques her interest. She moves her pen over the passage, but the pen leaves no mark. Instead, it copies the sentence and transmits it to a file on her office computer where she collects quotes worth saving. Before leaving for work, Sal types a code into her garage-door opener in order to locate its instruction manual. She had emailed the garage door manufacturer to ask for a new manual (she couldn't find her copy), and they replied explaining that, once she enters the code into the opener, "the missing manual will find itself."[25] She punches in the code and a beeping noise emits from a microchip that the manufacturer has glued into the manual to help their customers track it down whenever they misplace it. Sal's household products come imbued with an extra layer of care their creators have installed in anticipation of the user's needs.

In her car, Sal regularly sneaks glances at her "foreview mirror" displaying information on the windshield. Whereas the rearview mirror shows the usual reflection of cars behind her, the foreview mirror that Weiser has bestowed on Sal equips her with all the functionality that GPS navigation apps would later give smartphone-aided drivers. The foreview mirror notifies her of an upcoming traffic jam in time for her to change her route. It also highlights a new coffee shop that Sal just has to try, and she does. When she pulls into her office building, the foreview mirror has already scanned the parking lot for openings, having communicated with sensors in the asphalt to scout out the place in advance of Sal's ability to perceive it. She follows the turn-by-turn cues to the nearest available spot.

An employee ID badge clipped on her shirt pocket is the key to everything that happens next: "As she walks into the building, the machines in her office prepare to log her in."[26] The recognition of employee badges converts the building into a kind of metacomputer that accommodates each staff member's stated preferences as he or she nears. Sal's footsteps in the lobby boot up the electronics on her desk; her first step into her office acts as a password, granting her access to everything the moment she walks in. This, Weiser underscored, minimizes the idle moments Sal needs to spend staring at a PC. While her electronics are readying themselves, she swings by "the offices of four or five colleagues to exchange greetings and news."[27] Less time in front of screens, Weiser suggested, translates into more time socializing with colleagues.

Of course, the desktop isn't the only computer in Sal's office, and Weiser hardly mentioned it again. As Sal eyes the mass of clouds blowing in atop the Santa Cruz Mountains outside her office window, the window also displays the local forecast: "75 percent humidity and 40 percent chance of afternoon showers."[28] She can customize the window to map other datasets onto the setting, such as the stock prices of the corporations scattered among the valley or scores from sporting events held last night at any of Stanford's many stadiums poking out through the distant trees. A light by her door starts blinking. Any visitor might wonder what it means, but Sal knows and hurries down the hall. During her first day on the job, she programmed the light to blink the instant anyone brewed a fresh pot of coffee in the break room.

Weiser brings Sal back to her office to show readers one last application of tabs, pads, and boards—a feature that Xerox's marketing staff hoped to sell. Sal (presumably jumping out of her skin now from the gallon of coffee her author has fed her over the past hour) finally begins her work. She picks up a tab to activate the "virtual office" she is sharing with Joe, a design colleague based elsewhere. What follows feels like product placement tacked on at the end, perhaps in response to some executive's prodding. In any case, Sal has an unspecified quantity of tabs and pads spread across her desk, as does Joe on his. Each device hosts one of the various documents they are creating together; on her collection of little screens, Sal monitors the edits Joe is making. According to Weiser, "She feels more in touch with his work when noticing the displays change out of the corner of her eye."[29] When one of her tabs beeps, she takes it in her hand and motions at the Liveboard in the corner. Sal's gesture has sent the document on the tab to the Liveboard. It automatically enlarges the image of a paragraph that Joe is asking Sal to help him with, and she hears Joe's voice through the Liveboard's speakers. Sal gives it a quick read and again gestures with the tab, this time in the direction of a nearby pad, where she circles a word in Joe's paragraph that strikes her as being a little bit off. "I think it's this term '*ubiquitous*,'" she says to Joe. "It's just not in common enough use. . . . Can we rephrase the sentence to get rid of it?"[30]

Taking this playful jab at his own idea, on some level, fits with Weiser's self-deprecating tendencies. Weiser could've had Sal say anything here. She could have pointed out a grammatical error or offered generic praise—either would have conveyed the technology's capacity to support remote collaboration. Instead, Weiser chose to call into question the one word that was most synonymous with his budding reputation. And working this criticism into his hypothetical scenario may well have been Weiser's way to make light of real exchanges that festered unresolved in his mind.

Weiser's struggle to clarify his vision as he wrote the article had consumed him almost to a breaking point. It was a tall, fraught order for him to make others see what he saw in things that did not look like much yet. Already, Weiser couldn't help but dwell on the confusion ringing throughout the viral buzz over ubiquitous computing. He heard it in colleagues' suggestions to include flashy imagery in his article—such as the array of screens on Sal's desk as she collaborates with Joe—that didn't exactly follow from the philosophy he was setting forth. And then there were the new faces in his unconventional office hours at the Dutch Goose; they would stumble to formulate questions his article had already raised, or try to play devil's advocate by attacking him with points he readily agreed with. They'd say something to him and then comment among themselves, as if he were speaking on stage rather than sitting across the table. "It wasn't the kind of interaction you have where you sit down with a group of friends and ask each other about how your day has been or how your life is going," recalled Reich, who often sat beside Weiser during these sessions. "I could never tell[, were people] there because he was famous? Were people there because they liked him? Did they even know him?"[31] After a while, she stopped coming.

Weiser remained excited by all the interest his presence stirred up, not only during his routine appearances around town, but also at the traveling lectures he was being invited to deliver with sudden regularity at top universities and conferences. These presentations, it seemed, would give him plenty of chances to nail down the meaning of what he thought his writing was trying to say.

Cambridge, Massachusetts, was cloudy, windy, and cold as roughly six hundred spectators filed into MIT's Kresge Auditorium. Serious-looking students hurried past groups of businesspeople into the concert hall. Almost everyone in attendance had paid for the privilege to be counted as a Media Lab insider, pledging their support in the form of corporate sponsorship fees and conspicuous donations or, in the students' cases, tuition bills and countless hours lending research assistance for one of the lab's many projects. They took their seats, and a butler took the stage.

Nicholas Negroponte, the Media Lab's director, must have eyed the scene like a playwright witnessing a performance of his script.

Negroponte had given much thought to butlers during the past two decades. Whenever he graced interviewers, audiences, or readers with his predictions and prescriptions, he usually evoked the conduct of a well-trained English butler to personify his sense of how computers ought to behave. "I find that the most constructive way to think about computing in the home," he said several months prior, "is to think of human servants and what they did in the old days for the plutocracy of this world."[32] For the day's event—world's first symposium devoted to the topic of "interface agents"—his lab had hired a living visual aid to animate his thesis. The butler-dressed actor was meant to give the crowd a taste of how informational assistants could help them in their own lives. To christen this initiative, Negroponte and Pattie Maes, another star researcher at MIT, had solicited talks from a dozen experts, including a few living legends. But no one was more instrumental to the day's proceedings than the butler.

The actor playing the butler on stage had been, as the audience read in their pamphlets, "expertly programmed" to embody the interface agents that Negroponte, Maes, and their collaborators were coding into digital existence. The butler explained in his opening remarks what interface agents were and how they would help computer users handle the "deluge of information" that poured into their email inboxes and all the other messaging platforms bound to enliven their screens.[33] In technical terms, interface agents were "semi-autonomous computer programs that filter queries, suggest actions, and automate tasks for their human users," said the butler.[34] Computers of the future—desktops and especially handhelds—would be outfitted with a cadre of agents ready to scour troves of data on the user's behalf. To keep things simple, all these agents would be represented by a single, human-like figure (think Apple's Siri or Amazon's Alexa) that would converse with the user and relay his commands to the unnamed others. It was Negroponte's butler's job to be a theatrical expression of this human-like persona. He had been instructed to help the crowd make sense of the events happening on stage in ways that simulated how an interface agent might serve as a computer user's personal assistant on screen.

The butler's primary duties consisted of introducing each speaker and doing everything he could to underscore the key points of their talks for the audience. This included making exaggerated facial expressions toward the crowd, such as a look of surprise when a speaker shared an astonishing statistic, or a look of pleasure to accompany the speaker's descriptions of desirable technological outcomes. Occasionally, too, he would chime in during the presentations with explanatory commentary (and snide comments) spoken directly to the audience—much to the disdain of one presenter, as Weiser would make clear once his turn to speak came.[35]

The butler could only partially mimic an interface agent, however, since these software aides were also meant to behave in ways their appearances did not convey. Cloaked as obedient servants, interface agents operated like spies, too, closely monitoring the user's digital activities. Their value lay in their ability to instantly comb through databases (e.g., personal calendars, email inboxes, hard drives) in order to answer the user's questions and to execute requested tasks. In the same issue of *Scientific American* in which Weiser's article appeared, Negroponte as well as ex-PARC–turned–Apple researchers Alan Kay and Larry Tesler had each stressed the need for interface agents in their published musings on technology's future. Tesler, who led the development of Apple's Newton handheld device, asserted that computing would become largely mobile by the decade's end. Mobile users did not have a robust keyboard or mouse, and generally could not allocate the time or attention to scroll through lengthy documents on a tiny screen. Tesler's article proposed that speaking and listening to the handheld device would eventually become the best way for people to interact with it; these "pericomputers," in Tesler's estimation, would at their best serve as a lightweight extension of a desktop machine that granted continuous access to its files and data.[36] This extra access would be useful only if handhelds also equipped people with a quick, reliable way to find the information they were looking for. Interface agents seemed fit for the job; their unique brand of artificial intelligence spared the mobile user from clicking through a series of windows, menus, and options. You would simply talk to the device in your own language and the interface agent would deliver the relevant details it had culled from the heap. It knew where to look because it knew a lot about you from all your data.

Alan Kay, who was the first speaker the butler introduced that morning, believed that interface agents would make computers *intimate*—that is, more personable than the personal computers he had co-invented at PARC during the 1970s. He recalled the early brainstorming sessions of those years and likely surprised many in the audience when he said, "At Xerox PARC, we thought about doing an agent interface. But we realized . . . we had no idea how to do an agent interface really, really well. So we fell back on doing tools."[37] The concept of agent-oriented computing was indeed older than the desktop metaphor around which PARC designed the software for the Xerox Alto in 1973. Stanley Kubrick's classic 1968 film, *2001: A Space Odyssey* (itself based on a 1951 short story by Arthur C. Clarke), had cast the interface agent HAL as its sole antagonist. The film's astronauts rely on their conversations with HAL to control the spacecraft, though HAL handles most of the essential tasks through its own initiative, even before it turns on the crew toward the film's end. What enables HAL to kill off all but one of the astronauts is the system's total understanding of each crew member's behavior. HAL knows their routine and identifies the opportune moment to strike.

Setting aside its murderous streak, HAL still stood as a prime example for understanding how interface agents would work in everyday computing. The agent's capacity for intelligence, its ability to meaningfully answer its user's questions, was rooted in "learn[ing] a user's goals."[38] Pattie Maes explained this functionality in greater detail when the butler called her up to the stage. Interface agents were constantly observing on-screen events and retaining whatever they could. "The agent maintains a whole memory of examples," she said.[39] For instance, your agent would process every word in the emails you sent and received, every bit of data in all your files. Once processed, each minute detail would be as familiar to the agent as your own name is to you. It would know exactly all the content, along with any recurrent patterns that could be discerned from the history of your computer usage—including patterns too obscure for most humans to notice. Whenever the agent confronted a new situation, Maes continued, "it [would] retrieve all examples that closely match[ed] this new situation and make a prediction for which action to perform in this new situation based on the old situations it has memorized."[40] And so, however earnest and chipper the interface agent might present itself in your exchanges with it, at the base of its human-like mannerisms would be a relentless, omniscient calculation that churned out an ever more accurate and foretelling model of you. Every answer an interface agent offered, every action it initiated, would be premised upon its cumulative mastery of all the electronic moves you had heretofore made.

Only Negroponte dangled nakedly before his corporate sponsors the promise of leveraging all this user data to fuel targeted advertising campaigns. He had winked at such a prospect during his *Scientific American* article, in a paragraph extolling the many ways interface agents could enrich global marketing efforts: "Imagine how delighted General Motors and Nissan would be to have the opportunity to advertise to you specifically when you start looking for a new BMW."[41] Surveillance capitalism, according to the critic Shoshana Zuboff, would not be put into practice for another decade (when Google quietly began selling users' search data), but Negroponte appeared to be among the first technologists who suggested this approach.[42]

Weiser, meanwhile, sporting a rare tie and his trademark red suspenders, waited for his turn to address the crowd, aware that the other speakers were likely aligned against him. Weiser had picked fights with a few of them in his interview with reporters over the summer. Most pointedly, for example, he told a Swedish technology magazine shortly before the symposium: "I feel sick when I hear Alan Kay, Apple's research guru, talk of intimate data processing as the next step. Computers are a part of my life, like paper, pens, and chairs, but I don't want to become 'intimate' with them."[43] The list of people and concepts he publicly criticized did not yet include Negroponte or the notion of auctioning off users' personal data (though he had a few jabs planned for Negroponte during his presentation that afternoon). This dystopian prospect occupied a clear spot on Weiser's radar of looming disasters that pervasive mobile communication networks might unwittingly foster if computer scientists were not careful. He had mentioned, too briefly and too casually, in his *Scientific American* article that "marketing firms could make unpleasant use of the same information that makes invisible computers so convenient."[44] Weiser's reluctance to emphasize the magnitude of that unpleasantness stemmed from his confidence that cryptographic techniques and legal regulations would be developed to "safeguard private information" before the most nefarious scenarios could take root.[45] In mistaking privacy protections as an eventual given, he chose to focus instead on speaking out against mobile computing projects underway at Apple that were, akin to similar work at MIT, moving toward a future filled with interface agents.

The first version of Apple's new handheld device, the Newton, boasted no such interface agents when CEO John Sculley unveiled it earlier that year at the 1992 Consumer Electronics Show. Commenting on the Newton's debut and its designers' plans to incorporate an interface agent who would talk to its users, Weiser flatly dismissed it in the *Chicago Tribune*: "Newton isn't anything new. It's still one person, one computer with too much emphasis on the technology. Truly useful tools don't call attention to themselves."[46] His contempt for the Newton actually ran much deeper. The Apple handheld already represented everything that Weiser felt a mobile device shouldn't be. Like the Poqet PC, the current Newton promised to be little more than a watered-down desktop in a smaller package—it showed no ambitions of connecting with the objects in one's surroundings the way Weiser and his colleagues were using their tabs around PARC. But it was Apple's long-term plan of adding interface agents to improve the device—the next-generation Newtons ostensibly possessing some level of AI—that Weiser had decided to combat in his talk today.

Weiser thumbed through the presentation slides in his lap one last time. The butler called him up to the podium. Almost as soon as the audience read the title of Weiser's presentation on his opening slide—"Does Ubiquitous Computing Need Interface Agents?"—he supplied the answer. It did not, Weiser assured them, indicating that he disagreed with everyone who had spoken before him. But that was just the start of his answer. What he went on to say proved far more oppositional. He regarded his fellow panelists' eagerness to equip mobile devices with intimate AI personas as a threat to his own vision. Irrepressible traces of anger, awkwardness, and astonishment must have fluttered around the auditorium, as members of the audience looked to one another in search of a consensus reaction. Reveling in the tension, Weiser pointed across the stage at Negroponte's actor: "Our excellent butler here today actually intrudes into the proceedings," Weiser told the audience. "He's fun, but you really wouldn't want a computer like that."[47]

Software agents, Weiser warned, put "the interface in your face."[48] Those who had read Weiser's *Scientific American* article knew that disappearance was the quality he prized most in a technology. Having our access to computing mediated by interface agents stood to raise yet another barrier of virtual machinery, it seemed to Weiser—one that would further impede our tacit engagement with the world. "Personal computing," he continued, "is the wrong idea and *intimate* computing is even worse."[49] Even though agent-oriented mobile devices dispensed with the desktop model that Weiser critiqued, he argued that agents would perpetuate the PC-era habit of making digital devices "a single locus of information," which people might feel compelled to attend to constantly.[50] Having people chat with their own portable, talkative AI assistant would keep them focused on a computer, even in the absence of a keyboard, mouse, and monitor. The computer traveled with them under this model, but it would be speaking to them wherever they were—perhaps in the very way that Negroponte's butler spoke to the crowd during the presentations. Interface agents, too, would by virtue of their design be talking over the user's live encounters with other people and things, putting the user in a position of having to juggle multiple conversations at once. (For this reason, Weiser thought vocal interaction should be a last resort, to be employed sparingly and very briefly.) Moreover, because interface agents monitored and analyzed so much personal data in order to construct models of their users' psychology, agents could develop an increasingly keen ability to get their users' attention and keep it.

Ubiquitous computing, in contrast, called for a distributed network of unremarkable interfaces (tabs, pads, and boards being phase I) that together presented and organized bits of electronic information "by place, time and situation."[51] Abandoning the notion of an immersive "single locus"—be it a desktop you sat at or an interface agent you carried around—was a prerequisite in Weiser's eyes. So long as one screen or even one AI voice stood apart from other things in your environment, computing would never "vanish into the background" and effectively become "an invisible part of people's lives," as Weiser argued it should (knowing full well that his current ubicomp prototypes were not yet exemplary in this regard, either).[52] This distinction between single-locus systems and a network of interfaces built into an environment topped the list of contrasting design principles that Weiser put on the projector, in order to highlight the differences he saw between ubicomp projects at PARC and the interface agent initiatives at MIT, Apple, and elsewhere.

The last of these listed distinctions pointed to a deeper, existential concern. Interface agents carried the potential to reduce the user to the most reactionary of creatures. Employing an agent to automate tasks on your behalf would mean outsourcing your direct engagement with large swaths of possible stimuli, both virtual and physical. Like the rich person who uses a butler to screen incoming calls and to see who's knocking at the front door, the wielder of an agent-oriented mobile device might have an information servant handle all manner of cognitive and communicative duties. You interacted with the agent, and the agent reported back about its encounters with the wider world. The more you interacted with the agent, the less you interacted with the world—or, at the very least, conversing with your agent stood to chronically divide your attention. You might have to ask your agent

Some different design principles

Interface Agent	Ubiquitous Computing
single locus of information about me	distributed, partial information by place, time[,] and situation
command the computer	what computer?
personal, intimate, computer	personal, intimate[,] people
filtering	breathing, living, strolling
user interface	no boundary between you and machine
DWIM *do what I mean*	WIWYHIAFI *when I want your help I'll ask for it*
I interact with agent	I interact with the world

A reproduction of a table from Mark Weiser's October 1992 presentation "Does Ubiquitous Computing Need Interface Agents?" Courtesy of the Department of Special Collections, Stanford University Libraries, Stanford, California.

a series of follow-up questions to ascertain information you could've found quicker by performing the task on your own. A lot could go wrong in spoken exchanges, particularly when one party was a machine (as Suchman's ethnography had shown). Inviting people to converse with talking software was bound to yield the same frustrations that led desktop users to smack their monitors—and it clearly had in the case of MIT's Media Room project a decade earlier. Even when interface agents managed to serve users precisely as intended, that habit might weaken your inclination to think and act without the algorithmic blessings of your own little digital god. An organization or a society that entrusted interface agents with more tasks might come to place less trust in human intuition.

Weiser finished his talk with an example that he hoped would illustrate the philosophical differences that separated interface agents and ubiquitous computing. He asked the crowd to think about how these respective technological camps might go about the task of boosting a pilot's ability to safely fly a plane. First, he explained how technologists abiding by the butler metaphor would likely develop their interface agent to assist with the pilot's tasks in a manner akin to *2001*'s HAL. Serving as a kind of electronic copilot, Weiser imagined that the agent would "watch the plane's systems, advise [the pilot] about things that are going on, talk in [the pilot's] ear."[53] The agent would handle the taxing job of monitoring all the data from the cockpit's many instruments so that the pilot wouldn't necessarily need to. Instead of having to rely solely on their own readings of the instruments and their perceptions of the situation, human pilots could also look to the plane's interface agent for alerts and guidance. If the plane was veering too close to another aircraft's flight path, then the interface agent would sense this immediately from the data and would probably, Weiser surmised, tell the pilot something like, "Collision! Collision! Turn right and down!"[54] Assuming the agent got it right and the pilot heard the instruction correctly, the collision would be avoided. But the frequency of near crashes like this might increase if pilots grew accustomed to deferring to an agent. In any case, to the extent that pilots must loosen their tacit grip on the situation in order to attend to the agent, they stood to become less of a pilot and more of a person in a pilot's uniform who monitors a screen and executes its prompts.

Weiser agreed that the complexity of the cockpit, like the infoglut overwhelming PC users, was a problem that begged for new technologies. But rather than enlisting agents to navigate the data sprawl on our behalf, Weiser claimed, the ubiquitous computing approach went to the root of the matter. He told the crowd that the ubicomp solution would be to redesign the cockpit entirely. He would aim to bring the presentation of navigation information and the plane's metrics into better alignment with the pilot's natural gaze and the actions she routinely takes to fly the aircraft. The modern jetliner's complexity required pilots to oscillate between flying the plane and operating the various computational systems meant to help them make the best decisions. The design strained pilots as they endeavored to keep track of it all, but they needed to continuously read the instruments in order to inform their gaze at the horizon. Divvying up the job and having an interface agent monitor the cockpit data would effectively relieve pilots of the burden that generated their expertise.

The root of the problem, Weiser maintained, was the cockpit's inelegant organization; a more intuitive means of presenting data to the pilot could keep them informed without overwhelming them. Eschewing the notion of bringing AI into the cockpit, Weiser said: "The ubiquitous computing approach would be to present airspace information for continuous spatial awareness, as in everyday life—[the pilot] would no more run into another airplane than [he] would try to walk through a wall."[55] Weiser's guiding ideal, in other words, was to weave the data so tightly with the pilot's casual perception of flight events as they occurred, such that the computer-generated information could inform the pilot's thinking as naturally as a lightning strike he spotted in the distance. Rather than remaining arbitrarily arranged on a crowded dashboard, flight data would be presented to the pilot on the basis of "place, time[,] and situation," so as to more closely overlap with his live perception of the sky. Weiser offered no details as to how exactly a ubicomp cockpit might look—it was just a thought experiment, and he had no such project in the works. Perhaps it was just a placeholder, an idealized alternative that he could hold over competing technologies to expose their flaws. Ubicomp tended to function like that whenever he stretched the idea too far into the future. He seemed to have had in mind something like the displays featured in his "Sal" story: augmented-reality windows bearing data visualizations and electronic tracks relevant to the scene, and perhaps foreview mirrors giving advanced glimpses of information pertinent to upcoming stages in the planned route.

Following Weiser's talk, the last session of the symposium featured all the day's presenters in one long row on the stage. The butler and an additional moderator fielded the audience's questions, which dealt one after another with the technical challenges involved in building interface agents, as if their viability were the one remaining concern—as if Weiser's argument had already been forgotten. In their responses, the other speakers referenced Weiser only to note his dissonance from the prevailing tenor: "I think I heard coming from each speaker except Weiser that . . . ," began one response, for instance, from a Bay Area software engineer working on an interface agent named "Jeeves."[56] But one member of the panel, seated on Weiser's left, was evidently stewing on Weiser's criticisms. As the end of the session neared, after someone finally asked Weiser a question about ubicomp, Negroponte could not keep silent about Weiser's presentation any longer.

"To compare an agent to the copilot of an airplane is mind-boggling," Negroponte exclaimed.[57] He elaborated his objection on the basis of a distinction—the interface agents would be an expert about the pilot, not the airplane—that might have furrowed brows in the audience, had they not been so tickled by the academic drama their leader was treating them to. Because, of course, one would assume that the agent possessed commanding knowledge of the user (e.g., pilot) and the system it mediated on the user's behalf (e.g., airplane)—otherwise the agent would never become the intelligent, well-trained servant that Negroponte promised. Weiser waited for an opening in which to venture his response. Negroponte continued: "You are either the pilot because you enjoy flying, or you delegate flying—as most of us do most of the time—to an agent!"[58] As the butler rose to adjourn the session, it was clear that little time remained for a rebuttal. Weiser eked out a one-line defense, vaguely alluding to *2001: A Space Odyssey*: "Open the pod bay door, HAL. . . . You're dreaming, Nicholas," he quipped.[59] The crowd left with plenty to talk about as they exited into the dreary October twilight, and Weiser could guess at what they were saying.

Weiser's flight back to California the following afternoon was marred with a new sort of disappointment that he brought home with him. The illusion of an everlasting sunset—that sliver of amber still glowing for those flying west in the evening against time and dusk—passed outside his window. Given what he had said in the newspapers beforehand and all he had dared to say on stage, Weiser's chances of returning from MIT unscathed were slim. But the casual blowback, the maddening ease with which the other speakers had dismissed his carefully crafted intervention, was beyond what he expected. Having scarcely been heard hurt worse than not speaking at all. A year had gone by since his *Scientific American* article wowed the world with his ubicomp dream. The further he traveled from Xerox PARC, delivering all these talks elsewhere now, the more he struggled to convey the details that seemed to matter only to him and his closest collaborators.

Perhaps it was the tabs, pads, and boards. They had become the emblems of his vision, and images of them circulated quicker than did his explanations of their purpose. In photos, tabs could be easily mistaken for other tiny computers, like Apple's Newton. Tabs and pads were generally shown individually, as if they were stand-alone, "single-locus" devices. An image couldn't exactly capture the context-aware features that PARC researchers experienced by virtue of their in-house wireless network—you couldn't see how the content on a researcher's tab changed when he walked into a tab-filled room. Ubicomp was, at its core, this invisible dance of information, of software adapting to varied physical surroundings. The other handhelds hitting the market demonstrated hardly any concern for their user's immediate context. Still, the surface resemblance between these products and PARC's prototypes led people to lump them together.

For Weiser, context awareness was the overarching principle that ought to structure how applications operated on small devices. Under the desktop GUI paradigm, information and functions were grouped by file, icon, and window; extensive menus listed operations like "Save" or "Print." The onus was on the user to learn her way around the screen and navigate an array of programs by mouse and keyboard. Handhelds like the Poqet PC and the Newton were still structured along these lines. The contextual approach Weiser championed sought to incorporate knowledge of each user's surroundings, such that her movements through the setting brought the most relevant digital resources to the fore. Context—"place, time[,] and situation," as Weiser put it—promised to be the best driver for post-PC user experiences. Nimble systems designed to foster context awareness would spare the user from having to fiddle around with them like a PC—and from having to talk to an interface agent every time she wanted to access anything.

Weiser needed to create more distance between ubicomp and everything else. He needed to show more clearly how little his agenda held in common with mainstream handhelds. His lab needed to build new prototypes that showed why context awareness promised a more desirable future than artificial intelligence. Already Weiser had been calling tabs, pads, and boards "phase I" of the project. He now needed to figure out a second phase.

6

Retreat

IN ORDER TO HAVE A CHANCE AT making any kind of breakthrough, you always had to begin, Weiser seemed to believe, by transporting an assortment of PhDs to some unfamiliar venue for hours of faintly structured brainstorming. Whatever thoughts people brought with them were merely inspiration; the ideas that sprang forth afterward were the true beginnings of innovation. Usually, the meeting itself did not amount to much by the time it ended. Groups tossed around ideas, but, in Weiser's experience, they did not hatch new ones. Really, you gathered as a group for the sake of each individual there. Words spoken to the group reverberated against each participant's inner monologue. Gathering in an unusual setting somehow helped make the remarks stick. The afterlife of unpredictable exchanges between smart people was precious currency, and its true value was often realized weeks later, when some remembered fragment echoed in somebody's head while they were driving home or reading a book. A seemingly random aside uttered during such a session could later accrue into something worthwhile through the compound interest of persistent brooding.

Tabs, pads, and boards had emerged in the wake of the visioning meetings Weiser led five years earlier, when he and his fellow computer scientists met in an aquarium to imagine what computing could become after desktops and laptops. Now he wanted to convene the top researchers from multiple labs at PARC for a second off-site retreat to ask them another question: What's next for ubiquitous computing? The query wore its urgency lightly on the invitations Weiser sent out.

PARC was devoted to long-term projects, and Weiser had billed ubiquitous computing as a twenty-year quest, so there was no obvious need for panic. Yet, with tabs and pads in use around the building and Liveboards being sold to the public, Weiser already felt a course correction was in order. Technologists and journalists were fixating on tabs, pads, and boards without giving much thought to the broader philosophy within PARC that motivated their design. "The initial vision of ubiquitous computing," he wrote to his invitees in February 1993, "has been confused with other, more prosaic ideas: PDAs, mobile computers, even simple laptops."[1] The poetry that Weiser and his collaborators sought to infuse into their in-house prototypes—their sensitivity to the user's surroundings—had largely fallen on deaf ears. The more prosaic track was doing dangerously well in the zero-sum competition for research talent and corporate investment. To win the long game, you had to keep the vision alive and growing in R&D circles. If it failed to remain a hot topic in conference presentations and doctoral dissertations, then the window of opportunity might close before the technology was ready—the industry may well have moved on to the next idea or reverted to previous ones.

The mounting popularity around "ubiquitous computing" had it inching toward the brink of meaninglessness. The more the world talked about it, the further it strayed from the ideals Weiser's team had in mind. He now made it a point to tell the press that tabs, pads, and boards—the first phase of ubicomp research—would be over soon, at least as PARC was concerned. "Instead of a world of smaller and lighter computers, which is the current and most accepted trend, we are oriented toward what we call 'the computerized space,'" Weiser said.[2] He wanted, in other words, to stuff the gadgetry of ubicomp inside familiar objects instead of continuing to build new gadgets. Rather than create improved pads or tabs for users to carry around or attach to other things, he now hoped to create improved, digitally connected versions of things people already used. Tabs, for instance, would cease to exist, but their capabilities would be installed directly into coffeepots, conference rooms, lighting, office doors, the HVAC system, and anything else around PARC the computer scientists could add to their local wireless network. The first five years of pursuing his vision had taught him to be wary of stand-alone screens, no matter how small they were. If something could be even loosely called a "computer," then people would likely think it was akin to a desktop PC.

The rest of Silicon Valley saw little cause for concern in any of this. Technology executives were busy hunting for more screens they might convert into personal computing platforms. While Apple neared the release of its Newton and collaborated with MIT researchers on a pipeline of projects aiming to upgrade tiny computers with AI software, other big companies were quickly warming up to Jim Clark's "telecomputer." Clark's pitch to turn home television sets into online portals for instant messaging, movie streaming, news, and gaming had enticed Time Warner just a few months earlier to pay his company Silicon Graphics $30 million to build the machine on a two-year deadline. The competition hastily rang in the new year with announcements of their own for 1993: Viacom partnered with AT&T, Microsoft with Tele-Communications Inc., and Digital Equipment Corporation with US West Communications. All believed the interactive television might be the killer app that would finally bring interconnected computing to the masses. This frenzied race was, in Weiser's eyes, just another step down a long road heading toward virtual reality.

By April 1993, Weiser's proud insistence that ubicomp was "off the beaten path" was as true as ever.[3] Eager to galvanize its second phase, he had spent months planning an off-site brainstorming retreat for a select group of PARC researchers to be held at a Jesuit chapel deep in the woods of nearby Los Altos. Lucy Suchman, along with her fellow anthropologist Gitti Jordan, agreed to help Weiser lead the thirty-five participants through two days of discussion that would perhaps set the goals for their work over the next five years.

By the end of the second day, the optimism that Suchman had mustered for the retreat would fade completely. The gathering began on a promising note, in a rotunda-shaped room where Weiser made some opening remarks before a wall of windows that overlooked the valley's treetops and the San Francisco Bay beyond. He encouraged everyone to take a step back from the present and set aside everything that ubicomp had been so far. "Tabs, pads, and boards were always only an initial step," he said. "To take the next step, it is useful to return to the long-range core of ubiquitous computing."[4]

For Weiser, this meant circling back to his intellectual muses: Polanyi's notion of tacit knowledge, Heidegger's notion of entanglement, vague notions about anthropology, and perhaps some old sayings from the reevaluation counseling movement. Now that the ubicomp agenda was again a blank slate, those conceptual foundations were back at the center of his speeches. Each offered a way celebrating instances where technology felt like a carefully grafted outgrowth of our intuitive capabilities. Any instruments deserving to be called "profound technologies," according to the criteria Weiser derived from these sources, had managed to enhance the "two-way openness of us with the world around us."[5] White canes, hammers, and the like extended individual agency while also heightening our ability to feel our way through the social, material surrounds. Backyard toolsheds were full of profound technologies, but computer scientists had still not produced many. More than any other variable such as portability or size, Weiser underscored that ubicomp's distinguishing trait was the human-world relationship it aspired to orchestrate. Whereas virtual reality was about "faking the world" and AI-powered interface agents effectively "shielded [users] from the world," ubicomp remained rooted around "being in the world," said Weiser.[6] He concluded his opening remarks with a nod to the importance of anthropological perspectives and reiterated his hope that the days' workshops might initiate new collaboration between his lab and Suchman's research group.

This plea, both sudden and fairly belated, to bring PARC's social scientists into the fold of ubicomp development was still timely in a way. The anthropologists' study of the ground-operations staff at San Jose International Airport—which involved roughly four years of close observation, data collection, interviews, filmmaking, and scholarly write-ups about the project—had wrapped up by the spring of 1993. Their findings only deepened Suchman's commitment to examining and articulating the unacknowledged ingenuity that rank-and-file workers showed on the job. It was rewarding to draw attention to the little things that quietly kept an organization running. At the same time, the group was keen to see what more might come out of this process. At design conferences and in distant countries, scholars were advocating that workers ought to have a say in any technical developments aimed at their work; Norway had even passed a law in 1977 to make this mandatory.[7] While no such radicalism lingered in the minds of Silicon Valley executives, a joint project uniting Suchman's group and Weiser's lab could be a decent starting point. The Computer Science Lab was, despite its cloistered tendencies, one of the most famed R&D sites in tech, and ubicomp was still garnering much support from Xerox and heavy interest elsewhere. A collaborative undertaking would provide a platform to show the tech world at large how much it could learn and benefit from paying better attention to the people whose lives they aspired to make easier.

Suchman and Jordan came to the retreat ready to make the case for a genuine, reciprocal partnership; their team's ongoing research was too interesting to divert for anything less. After Weiser's presentation, a quick overview of the current mobile-hardware market followed, and then it was time for the invited participants to share their visions. Each had been asked to write a position paper in which they either described a setting where adding ubicomp technology might help, outlined a ubicomp artifact that would aid their own work, or pitched a ubicomp device that could be built in five years. Several papers in, it became wildly apparent how differently Suchman's team worked: they presented a honed message that cohered from one group member's presentation to the next, as if they had actually met to discuss their ideas beforehand. Talks given by the other participants weren't like that. An alarming number of them started off with the admission that they had come simply "to find out what ubiquitous computing" was.[8] Among the other participants—the ones who had done the assigned reading—

two opposing camps emerged: technologists who still assumed ubicomp meant building cheaper, smaller versions of the PC as they knew it, and those who understood that Weiser had no such intention.

The latter contingent floated some intriguing if disparate proposals. There was a "Dick Tracy Watch" that would allow its wearers to share their current location, remotely control their home's lights and door locks, monitor their health metrics, and receive audible navigation directives.[9] There was a "Star Trek tricorder" that might "probe the surrounding electro-mechanical subsystems" to diagnose malfunctioning electrical systems.[10] Bucking the sci-fi trend, one paper reimagined the traditional children's lunch box. The futuristic ubicomp version would display a live video of the kid's parent on the interior panel when she opened the lunch box. The parent would, in turn, watch the child eat and "whisper words of love and encouragement (or reproval if the child [was] not eating correctly)."[11] This was Rich Gold's pitch, of course; no one else had his knack for conjuring the mundane realities into which every hyped technology eventually settled. His portraits of the future were sincere, yet ambiguous; playful, but not sarcastic. Elaborating on the parent-child video calls supported by his envisioned lunch box, Gold continued: "Questions can be asked, endearments can be exchanged. They will be a family together, despite spatial/temporal separations. . . . I cannot even begin to tell you how big the market for the Home-Away-From-Home Lunch Box will be."[12] You never knew just how Gold felt about the scenarios he crafted—they were like parables more than pitches, and they revolved around the desires a technology seemed to be rooted in, mentioning its features only in passing.

The rest of the computer scientists did not skimp on technical specifications. One after another, they unveiled the name of their hypothetical device, waved obligingly at a possible use case or two, then proceeded into a lengthy analysis of the engineering challenges the device would entail. Still, if you squinted past that and pieced together the collective gist of their ideas, you could see a veritable catalog of connected, computerized objects that didn't at all look or feel like computers. From the lunch box to the watch, the scientists had made it a point to imagine building into everyday things a set of digital capacities that appeared tailored to suit the particular contexts in which people might use them. These sketches aligned well enough with Weiser's sense of ubicomp's second phase, and they confirmed his notion that PARC could progress beyond tabs, pads, and boards.

On the other hand, from a process standpoint, the whole exercise illustrated with almost laughable accuracy the prevailing model of innovation that Suchman and Jordan hoped to challenge ("Design from fantasy," Jordan would call it in her post-workshop evaluation[13]). Their position papers opened with deft assessments of ubicomp's conceptual basis. Suchman saw Weiser's work as striving for "the disaggregation of personal computing from the workstation," whereby digital technology would be regrouped around the "activities of their use" rather than all lumped together in a single box.[14] And this push to make computing power "uniformly available in a particular space," Jordan added, implied that interface designers' tendency to map human-computer interactions from an individual user's point of view was no longer sufficient.[15] The ubicomp designer would need to ask "how computing [could] support group interactions," and doing so would require nuanced investigations into the tacit knowledge that groups of workers cultivated on the job (at workplaces other than just PARC).[16] The tabs, pads, and boards being prototyped around the Computer Science Lab were still privileging the vantage point of the individual user, as Weiser had done in his *Scientific American* vignette about Sal's morning routine.

The researchers in Suchman's group—Work Practice and Technology—had a proposal of their own in mind. Their ethnographic study at the airport and their familiarity with related efforts at other sites informed Suchman's suggestion that "centers of coordination"—airline control rooms, 9-1-1 dispatch centers, or air traffic towers—could be fitting venues for ubicomp research and development. "All of these settings share a concern with coordinating the actions of people and the deployment of equipment through time and across space," she said.[17] All were, by the same token, profoundly ill suited to the solutions that desktop PC's offered to more humdrum workplaces—which was also to say that such contexts differed wildly from the offices where computer scientists worked at PARC, which furnished their attendant presumptions about the nature of work. The San Jose Airport staff, for example, relied on a delicate balance of phones, walkie-talkies, video screens, electronic databases, whiteboards, and paper documents to quickly relay precise information from the operations room, ground crews, and airline headquarters in Dallas. Prior to each flight's departure, the amount of luggage and fuel and the number of passages got jotted down on clipboards, shouted out over radios, typed into computers, and confirmed on phones—all in order to determine the correct flap settings for the aircraft's weight. Surely, workplaces of this sort would provide an exciting milieu in which to reimagine ubicomp. Compared to the typical office, a center of coordination promised a more dramatic space for exploring just how unique and practical an ensemble of a hundred interconnected, computerized objects might become. Rather than filming computer scientists lounging around with tabs, pads, and boards during a team meeting (as PARC's 1991 ubicomp promotional video had done), you could portray more-dynamic scenarios, featuring airport crews or first responders, perhaps, that showcased teams of mobile professionals making use of electronic data and digital tools meant to inform their work on airport runways or in busy city streets.

What Suchman and Jordan were also hinting at was another kind of creative process that, with their guidance, stood to breathe new life into ubicomp's applications. The airport was just one possibility among several that an interdisciplinary mix of the retreat's participants might embark on together over the coming years. The anthropologists could forge the relationships with real staff at a relevant job site, where they and the computer scientists could then exchange ideas with prospective users. These users—these *people*—wouldn't simply be test subjects like rats in a maze. Through ethnographic observations and dialogues about existing technologies in that particular workplace, a cross-lab project would "involve members of the community in problem formation, design, prototyping, and the 'growing' of new technologies."[18]

Suchman's computer science collaborator Randy Trigg (who had been a student of Weiser's back at Maryland) explained to the skeptics in the room why he enjoyed co-constructing technology with the people he was developing it for.[19] He spoke to the magic of inductive learning that happened when he visited worksites with Suchman—how months of careful immersion yielded surprising details that had the power to make or break a technology's potential to enhance the well-being and productivity of a group of workers. An additional benefit of pursuing invention in this fashion, Trigg attested, was how specific insights gleaned from observing one workplace might carry over to other scenarios. Hence, while you might set out to create database software for a doctor's office, the solutions you designed into that project might also be adapted to serve the needs of a hardware store. The more site-specific projects you worked on, the larger your inner library of knowledge about particular work practices became. Keenly understanding many different contexts put you in a better position to design for various cross sections of the broader population.

This unconventional image of a technologist out in the field didn't cast favorable light on traditional methods. Silicon Valley technologists typically stuck to their labs and just as typically imagined more or less everyone as their intended user. When held against ethnographically informed methods, however, it was threateningly plausible that the technologist's picture of "everyone" was just an amalgamation of their own work experiences blended with images of actors playing other kinds of workers in movies they'd seen. And yet, the history of computing and the computer scientist's high place within it made alternative models easy for them to shrug off if they felt like it. (They also knew that whatever they built would, given their hard-won technical expertise, prove better than anything a nontechnical person might try to build without them.) A renowned figure in PARC's user-interface group spoke up on behalf of most technologists at the workshop when he insisted, "User studies can be valuable[,] but they are not magic. . . . Technology push is responsible for as much advance as deliberate market pull."[20] Sometimes a scientist's or engineer's highly educated gut instincts, in other words, generated valuable new directions in R&D that no prospective customer would be able to foresee. Of course, Suchman's group did a lot more than "user studies," but technologists still tended to reduce anthropology to this phrasing.

For all his praising of anthropological frameworks, Weiser, too, remained wedded to laboratory life. Two years before the present retreat, he had made a note to himself during a presentation Trigg had given at PARC. "Orienting prototypes towards users' work," Weiser wrote in his notebook, "has the problem of bringing nothing new."[21] A visiting anthropologist who observed the Computer Scientist Lab around this time later said of the ubicomp teams that all their favorite questions began with "What if . . ."[22] They maintained that radical inventions could be divined only upon a smooth plane of thought set apart from the dulling influence of quotidian norms.

Suchman had already grown to expect reactions of this sort. As the retreat wore on, through the breakout sessions and the meals and the evening walks in the forest behind the Jesuit chapel, one knew what to make of the technologists' ponderous nods and their noncommittal *hmmms*. Suchman recalled, when thinking back on that retreat much later, "My main memory is of how interesting it was, to me, how uninteresting the [San Jose Airport study] was to Mark and the others. . . . That was not the kind of 'What's Next for Ubiquitous Computing' that they wanted to think about."[23] Weiser wanted anthropological ideas to inflect how his lab imagined the everyday situations they would infuse with ubicomp, but he still loved to stand before a whiteboard in the cerebral quiet of a closed-off room. Among some of the other engineers, a hasty calculation of the anthropologists' political orientation fueled unease. One had once muttered that Suchman was a Marxist, without venturing any explanation as to why. In truth, she had never felt much compelled to read Marx.

What she was, she realized a bit to her own surprise, not long before the retreat, was a feminist.[24] Later that year, Suchman wrote an influential scholarly article that furthered the model of innovation she and her colleagues had outlined to Weiser's workshop attendees. The article leaned firmly on leading feminist thinkers such as Jane Flax, Judy Wajcman, and most of all Donna Haraway, whom Suchman had first met during a panel session they both participated in at the University of California, Santa Cruz. The work of these scholars furnished Suchman with some theoretical building blocks that she assembled into a framework for spelling out three contrasting approaches to R&D in the tech industry. She labeled them "design from nowhere," "detached engagement," and "located accountability."[25] "Design from nowhere" was, in short, the Computer Science Lab's tradition. One could imagine scenes from that ubicomp workshop replaying in Suchman's mind as she marshaled phrases to capture the essence of this mentality. Technologists working in this vein, she wrote, "problematize[d] the world in such a way as to make themselves indispensable to it and then discuss their obligation to intervene, in order to deliver technological solutions to equally decontextualized and consequently unlocatable 'users.'"[26] "Detached engagement" wasn't much better, and it was the model that Weiser seemed to be aspiring toward. Under this approach, Suchman claimed, a lab "[provided] distance from practicalities that must eventually be faced"—but facing up to those practicalities was left up to staff in some other department.[27] The technical class fielded directives from marketing execs and corporate leadership, as well as cross-disciplinary insights from other researchers, but firms organized in this manner diligently insulated their computing experts from worldly demands. The result was an R&D division that felt like "an elaborate social world within which one can be deeply engaged, but which remains largely self-referential, cut off from others who might seriously challenge aspects of the community's practice," Suchman wrote.[28]

Suchman's theory of "located accountability" stipulated that our knowledge was always partial and rooted in particular contexts: "Our vision of the world is always a vision from somewhere."[29] Designers in this camp thus took it as their responsibility to embed themselves in the very environments where their inventions would be used. Doing so meant amassing a venerable collage of qualitative research that faithfully portrayed the partial experiences and collective knowledge of as many constituents as possible. As Suchman now saw it, the ultimate output of technological innovation ought no longer be understood as "a single technology that subsumes all others," such as PARC's iconic graphical user interface or the computer mouse.[30] None of these inventions, by their sheer presence alone, wrought the revolutions they eventually came to be associated with. They were instead like a pivotal piece of wood added just so to an already-burning fire. Engineers could bequeath certain properties and specifications on a particular technology, but its fate hinged upon the meaning it garnered outside the lab. Genuine innovation was "the cultural production of new forms of material practices," and not simply a parade of shiny objects touted to render less shiny ones obsolete.[31]

Whereas corporate technologists considered themselves to be in the business of replacing existing gadgets with newer and better ones, Suchman preferred "artful integration." An emerging technology should be evaluated by what it contributed to "specific ecologies of devices and working practices," she suggested, rather than how thoroughly it portended to outdo established systems.[32] Suchman rested her case on the logic that located accountability stood as the most savvy and profitable stance for technology firms to adopt. Learning from users throughout a project's life cycle might require more up-front investment, but it would mitigate against the isolated engineer's propensity for spectacular flops, so many of which had been swept under the rug of Silicon Valley lure. Most important to Suchman, located accountability was a starting point for making the technocratic mechanisms of American technological production a bit more democratic. She and her team were forging ahead without Weiser's lab, and their next project would take them to a local law firm.

Weiser, meanwhile, had been deep into his own reading of feminist philosophy. He was enthralled by a recent book by Sandra Harding called *Whose Science? Whose Knowledge? Thinking from Women's Lives* (1991). Of special interest were Harding's remarks on "strong objectivity," a concept that implied, Weiser noted to colleagues, "that one cannot be objective without thinking about the social situation of one's work."[33] Weiser's fascination with Harding appeared to parallel Suchman's embrace of Haraway to such an extent that a reporter who interviewed each of them during a 1993 visit to PARC noted their overlapping interests with surprise.[34] Yet, it was easy to see why Weiser gravitated toward Harding's pages, which he cited in a separate effort to rethink how PARC might pursue innovation. At his insistence, a group of his fellow lab managers around the building agreed to read Harding's book. When one of them doubted its value, Weiser rebutted him with a vengeance. At minimum, Weiser wanted his colleagues to glean from the book how influential a scientist's life and culture were to the pursuit of scientific truths. A researcher's standpoint, partly inherited and partly cultivated, quietly factored into the problems she found most important and the kinds of questions that popped into her head. And it was the headspace of the lone scientist that Weiser loved to probe. In addition to stressing the urgency for more gender and racial diversity in hiring—"new science is enabled by increasing cultural diversity"—Weiser asked that the lab managers try implementing Harding's standpoint theory in project meetings with their research staff.[35] They should all take time, he urged, to reflect about why each was asking "these particular questions in this particular way"; furthermore, they might develop "a methodology of expanding [their] kind of questioning by other views and other lives. . . . This might cause [them] to change the questions [they were] asking."[36] Such was the intent of the methods Suchman had proposed and practiced, but Weiser opted for an interpretation that strove to invite the wider world into the lab, and at the same time reduce it to a thought experiment, so that the scientists might claim to grasp it without having to venture elsewhere.

PARC's buildings were still Weiser's favorite canvas, gallery, and time machine. Within them, he and his technical collaborators wielded homemade technologies that enabled them to approximate the infrastructure of their dreams closer than any other research facility or workplace could feasibly support, at least for the time being. And though the ubicomp retreat had not generated everything he hoped it might, Weiser now had more than an inkling of how the ubicomp experiment should evolve as he reread everyone's position papers in the incubatory calm of his office.

Even as Weiser continued to rally computer scientists around his push to define ubicomp's next phase, his misgivings about their progress plagued him for the rest of 1993. He concealed the worst of it as best he could. All the tabs, pads, and boards in use at PARC were bittersweet reminders of the gap between what they had built and what he now wanted to build. "It's awful," Weiser confided to a visiting artist in his office. "It's frustrating because the work goes so slow."[37] During this conversation, he also remarked that Suchman and Jordan were "not happy" about the direction of ubicomp. In his lab notebook-turned-diary, he had lamented, "[The Computer Science Lab] is broken. . . . We are too near-term. . . . We are far short of Alto-like success—not creating working models of [the] new paradigm."[38] He even wrote, "Have the courage to end this project," though it's unclear whether this was his own thought or a colleague's suggestion he recorded. Other pages bore reassuring scribbles tucked into the margins alongside meeting notes. "Notice fear, don't act from fear. (You're OK)," he jotted. "Stay in the moment. (I like it when you do that)."[39] A pair of tie-dye sweatpants made their way into his workday wardrobe. His beard grew bushier. Around this time, he decided that walking *around* sofas in the common area near his office was no longer necessary. A visiting reporter observed the habit during an interview with him: "Weiser stepped over the sofa's backrest, walked over the cushions, then stepped to the floor—meanwhile continuing to talk, as if climbing over furniture wasn't the least bit unusual."[40] For all the journalists, Weiser kept up his usual high-spirited persona. They printed the sound bites he knew they loved to lead with, like his go-to line "We're trying to undo what PARC did 20 years ago" (a provocative way to punch up his critique of the personal computing paradigm his predecessors had launched).[41] As he disparaged privately about his vision—which the press was still touting as tech's future—he committed to a next step.

There was a device being tested around PARC that Weiser's earlier presentations and essays had not said much about. It was called the "active badge," and Roy Want had designed it back in Cambridge, England, shortly before Weiser recruited him in 1991. Want's badge, roughly the size of a floppy disk, clipped easily onto a belt or a shirt pocket, where it would silently broadcast the wearer's location over a local network. Before mobile phones became popular, Want had initially pitched his badge as a better alternative to the pager (or "beeper") that a doctor would wear in order to remain reachable. Hospital staff would call the doctor's pager, and the pager would display a callback number, which the doctor would then dial on any nearby phone. Want's badge promised a more efficient system. Instead of phoning the doctor, the doctor's active badge would leave a trail of electronic signals that showed her staff which room she happened to be in at any given moment. Calls for that doctor could then be directed to the nearest phone, thus eliminating the lag of back-and-forth messages that pagers created.

Want and Weiser quickly realized that these badges might be used for much more than routing phone calls. The location data streaming from each badge were like notes strummed from a single guitar; from many badges, multiple data streams could be orchestrated into an array of songs. Telephone routing was the simplest end of the spectrum. They determined to make their lab the first test bed for an experiment that, they hoped, might soon be amplified throughout the entire building. Beyond allowing people to track their colleagues' whereabouts, a badge could also be made to convey its wearer's desires to sensors embedded in the rooms he inhabited. A sensor-laden room understood the wearer—the part of him that could be boiled down to data on a microchip—without his having to press any buttons or issue a voice command. Such environments instantly shared the data they gathered from the badge with whatever physical system they had been programmed to run. A speaker walking on stage in the auditorium might, for instance, trigger the stage lights to dim, activate the projector, and, if his badge was synced with his personal calendar, open up the slides for his presentation. The possibilities raised by badges and sensors, conjured in passing during the April retreat, hardened into the foundation for ubicomp's second phase by summer.

This latest initiative—the Responsive Environments project—sought to equip PARC's office infrastructure with traces of empathy, care, and possibly wisdom. It would necessitate many ongoing calculations. For starters, between the badges worn on scientists' shirts and sensors attached to the building's electrical grid, more than eight hundred variables pumped data continuously across the network.[42] The project's chief objective was to see how much the badges might reduce PARC's environmental footprint while also maximizing everyone's comfort. Those goals would've been mutually exclusive within the bounds of conventional energy-control systems. But because badges communicated the meaning of human movement so precisely with nearby objects, the room could deliver conditions tailor made for an occupant's present activity, all while ruthlessly conserving those forms of energy that were inessential. Weiser's office lights dimmed the moment he booted up his computer; they brightened back up when he logged off. His keyboard, like those in his colleagues' offices, was one of the eight hundred variables that the project monitored. The sun was another. Lights also adjusted themselves in accordance with the amount of natural light gleaming through each room's windows, as did the blinds, which opened and closed with the sun to regulate indoor temperature.

The ecological benefits to be gained from making offices responsive in this manner drove most of the project's proponents, including its technical lead, Scott Elrod. For Weiser, it was also something else. Tinkering with lights and thermostats was a first step toward a more tacit way of life. The beneficiaries of this long technological journey wouldn't even have to read Polanyi or Heidegger to feel warmly entangled with the world. Weiser's stance toward responsive environments was one of existential wonder—a sensibility he shared with his friend and collaborator Rich Gold. "When I walk into my office at PARC," Gold wrote, "I have become part of the furnace."[43] As he listened to Weiser talk about ubicomp, Gold mused that the fleeting joy we felt by linking ourselves with our surroundings, even in such rudimentary ways, stemmed from a longing to overcome the malaise of modern alienation. Gold believed "ubicomp's hidden, underlying assumption" was that "we find ourselves unhappily alone in an uncaring, dead universe. . . . We have 'rationalized' the world (and even ourselves!) through scientific reductionism into inanimate machines."[44] Read through Gold's eyes, Weiser's ambivalence toward desktop PCs seemed to have roots that ran much deeper than computing. But new kinds of technology, they both hoped, promised a more integrated sense of being. Ubicomp might reanimate things by giving them dynamic characteristics that people could recognize and engage with almost naturally. The Responsive Environments project led Gold to draw fawning parallels between its far-term possibilities and "the world of the Native Americans, where every tree, rock, cloud, hut, bead, feather, and buffalo contained within it a living spirit."[45] Gold's digital animism resonated with Weiser's attraction to the breakdown of subject-object boundaries advanced in the pages of his favorite European philosophy.

Compared to a standard ID badge, an active badge harbored a digital life force that extended its conventional function. Both sorts of badges displayed the wearer's name in printed letters for human eyes. The active badge stayed true to its role as an identification tool, yet it increased both the amount of information that a badge could communicate and the reach of that communication. Any physical system could, by virtue of badge-reading sensors, react to any badge wearer's presence as if it recognized him like a friend. And these sensors' tracking capabilities enabled the computer scientists to see through walls, essentially—they could spot one another from a distance just by glancing at a nearby computer screen, where a program called Bird Dog showed a real-time map of the building with every badge wearer's headshot marking his current location.

The badge was a bearer of ubicomp design principles that Weiser and Gold fancied applying to all sorts of other everyday things. They had fun imagining features for a "Ubi-Pipe," and Gold smuggled whimsical remarks on its hypothetical design into an article for a no-nonsense technical publication. The Ubi-Pipe would have a tiny speaker and microphone to support phone calls, owing to the pipe's "close proximity to the mouth." One could smoke a pipe while talking on it "without straining social convention," unlike speaking into a wristwatch. The Ubi-Pipe could be useful during presentations, giving pipe-touting lecturers the ability to point it at a screen and click on icons, just as they would otherwise do with a computer mouse ("pipes traditionally have been used effectively in this theatrical manner"). Finally, more to the point, Gold described how a Ubi-Pipe might feature little computational cues meant to inform the smoker during the act of smoking: "Detecting legal and illegal areas of smoking seems only right; while monitoring vital medical signs seems a proper inversion." Adding digital connectivity to a pipe or a badge—and doing so without noticeably altering the object's traditional appearance—felt closer than any previous ubicomp prototype to the sense of disappearance that Weiser longed to achieve.[46]

The active badge and the imaginary projects it brought within reach pointed Weiser toward a new metaphor, which he expressed in perhaps the most literary sentence computer science discourse had ever known. "Our computers," Weiser told his fellow engineers, "should be like our childhood: an invisible foundation that is quickly forgotten but always with us, and effortlessly used throughout our lives."[47] The power of this invisible foundation and its attendant glee would spread only if more researchers took to wearing the badge. Early adopters in the Computer Science Lab made up a fairly small percentage of PARC's total staff. Much of the building remained unrepresented within the growing suite of badge-oriented systems, and the energy savings won by the Responsive Environments project's initial rollout were just a fraction of the payoff to be had if everyone's office received this upgrade. The badge first needed to scale throughout PARC in order to prove the concept was ready for the wider world.

Weiser found a willing ambassador in John Seely Brown. PARC's director was keen to the broader potential of active badges. In remarks to the press, Brown favorably contrasted this new initiative with the "expert systems" framework that had long dominated AI research ever since he came to Silicon Valley. "Up to now, we have built computers centering on the 'thought' function of the brain, and we have ignored the aspect of vision, hearing, etc.," Brown said.[48] Badges and sensors gave a network the capacity to perceive live events in the physical world. The prior generation of AI researchers assumed that computers could process only elaborate simulations of things and not the actual things themselves. To leverage the machine's intelligence, they fed it a strict diet of rules and inferences meant to distill a virtual essence from unruly realities. Brown saw in badges the crest of a new wave. The representational cognitive mapping that had anchored AI in the 1980s might be rendered tedious and even sloppy. "The real world will participate in the representation," he wagered. "Compared to the previous approach, it is a huge epistemological shift—it comes down to abandoning the old Cartesian body/mind duality."[49] And it was with this sense of horizons unfolding that Brown wore his badge around PARC's hallways, until one afternoon when Suchman pulled him aside.

Brown had, as he recalled the scene, been walking around Suchman's end of the building with an active badge on his shirt pocket. The badges didn't work in the area where her team's offices were. This was by design, a bit to Weiser's frustration. The Work Practice and Technology group had declined his request that they wear badges and participate in the Responsive Environments project. Suchman didn't make a habit of confronting every badge-sporting researcher she saw (a majority of PARC scientists by now); but Brown was different. He was the head of the place, and here he was proudly donning a badge in their space of resistance. He remembered Suchman asking him point-blank: "Are you explicitly making a statement to us, or just implicitly making a statement to us? Because you seem clueless you are sending a signal."[50] The question had indeed caught him off guard. He knew, of course, about the general privacy concerns around the badges and readily conceded their validity. A flurry of recent newspaper articles had reported on the tracking experiments happening at PARC and a few other labs. Their authors speculated about the Big Brother scenarios that these technologists might spring on the rest of society. PARC, at least, had no such plans; transparent user controls and safeguards against data abuses were a shared priority. For example, Elrod's team had hardwired a special "smart/dumb switch" in the offices of staff who were participating in the Responsive Environments project. Flipping the switch to "dumb" rendered the room's sensors completely inactive and put a halt to all data collection. "One principle we go by here," Weiser told *Wired* magazine, "is to maintain individual control over who else sees anything about us."[51] Another principle was data reciprocity: a person had to share their own location in order to track other people's locations.[52] The Cambridge-based Olivetti Research Lab, where Want had first deployed the badges, exhibited less inhibition. Staffed entirely of engineers, Olivetti saw no issues with broadcasting each of its employee's location for anyone on the internet to monitor.[53] PARC's leadership considered their center to be at the ethical forefront of the industry. Brown supported a researcher's right to opt out of the badges.

But Suchman had additional reasons for asking Brown about his badge. The problem of deeper concern to her could not be resolved by technical safeguards. Sure, one could opt out of wearing a badge, but that gesture took on new meaning once the boss decided to opt in. By wearing the badge—in particular around the Work Practice and Technology offices—Brown was ostensibly endorsing the system that Suchman's group was intent on questioning, and his endorsement could undermine their stance. While PARC's computer scientists readily conceded that location sharing involved some privacy sacrifices, they failed to anticipate how organizational dynamics could exert pressure on those who refused to be tracked. This was palpable in sound bites Want and Weiser offered to reporters who visited PARC to get a closer look at the active-badge system in use. "It is in your interest as a professional to stay in touch with your colleagues," Want told the *New York Times*, as if that settled the matter.[54] Weiser insisted that the badges were intended solely to help employees "feel that they're in a community" whenever they were at work, even when stuck in a cubicle or enclosed in their office.[55] Many who wore the badges around PARC did attest to enjoying a heightened sense of collegiality, as well as productivity gains hastened by their ability to find people quickly.

A descent into the more alarming second-order effects might be less likely at PARC than in workplaces like Olivetti, but it was easy for social scientists to imagine how badges could alter labor conditions for the worse just about anywhere they were adopted by a critical mass. For instance, the same software that might save workers time by tracking a collaborator's whereabouts on a digital map also gave their managers a precise means of tracking employee work habits. The system could be said to deliver added value to all parties, but those lower on the hierarchy stood to lose in the bargain whatever autonomy they enjoyed before. Once this brand of pervasive tracking proved to be an effective means of monitoring productivity and even instigating it, how long would it take before the boss decided to utilize badge data in staff evaluations? How would an employee's refusal to be tracked be regarded then? At that point, wearing a badge would likely go from being "in your interest as a professional" to being a basic expectation.

Experimental applications that some computer scientists were building to boost the active badges' functionality at PARC unwittingly played into this scenario. Two such programs, Video Diary and PEPYS, generated video and text records documenting every event during a badge wearer's workday. Both systems categorized the nature of the worker's activity by accounting for the location where it took place and the identity of other badge wearers who were present. The stated payoff motivating each system's design was to give workers a better handle on their own performance; however, again, what the workers stood to lose from participating in these mechanisms dwarfed whatever they might gain. And what about the enhanced sense of community that location sharing created? This, too, entailed bittersweet trade-offs that tended toward the bitter. The initial novelty of constant connection might develop, like an old boys' network, into a casual professional circle that tacitly privileged the badge-wearing in a crowd and excluded those whose photos didn't appear on the screen's map. Building this augmented sense of community among badge wearers might thus fracture the community at large. Membership in the former meant pledging away your right to complete solitude, save for the moments in your office where you dared to flip the switch from "smart" to "dumb."

As the debate on privacy grew louder around PARC and in news articles about the badges, Suchman began to ponder a subtler concern that no one was discussing. It came to her in a flash when she heard the computer scientist Marvin Theimer wax gleeful about a new application he was working toward.[56] The convenience of having any PARC computer instantly configure itself to him had inspired Theimer to explore how such badge-enabled customization might extend to other, more foreign settings. The prospect of immediate, ubiquitous translation excited him most. He asked colleagues and reporters to imagine themselves visiting Japan, walking up to a Xerox copy machine, and being pleasantly surprised to see the copier's instructions automatically switch from Japanese to English. He could even foresee a day when entire towns and cities might accommodate each strolling tourist's preferred language: "The street signs around me [will] change to English as I walk by."[57] Most engineers delighted at this prospect; they marveled at the thought of stripping international travel of its logistical snags.

To Suchman, omnipresent translation wasn't just another neat app in the expanding array of functions the badges made feasible. It placed within arm's length the vaguely irksome thing she could never put her finger on before. In Theimer's fantasy—wherein every foreign language yielded to a badge wearer's native tongue—Suchman saw the utopia that some ubicomp researchers seemed to be chasing all along: "a world which is always familiar."[58] Between embedded machines that adjusted a room to accommodate each user's preferences and urban locales that redefined themselves according to each passerby's frame of reference, Suchman realized a common thread linking various projects was "this desire to never feel that you are out of place."[59] That desire was antithetical to anthropological sensibilities. An ethnographer who set out to do fieldwork among any group she wished to study always expected to feel perplexed. If you thought you understood everyone shortly after arriving on the scene, then you were doing it wrong—your initial assessments of the group's culture were likely marred by imported norms and concepts that you imposed upon the scene. Ethnographic methods aimed to prevent anthropologists from clinging to familiar reference points as they sought immersion in something different. They spent many months, often years, participating in the communities they researched in order to learn from the inside out, constantly guarding against the impulse to revert to assumptions, their own biases, their own values, their own terms.

When held against the intellectual standards of Suchman's discipline, that impulse to personalize the physical environment by digital means now seemed like some weird new breed of colonization. Futuristic ubicomp scenarios—from Theimer's letter-shifting city signs to Weiser's digital footprints showing traces of pedestrians by Sal's house—aimed to keep you always in the know. Maintaining this stance, or at least the semblance of it, would invariably require the technology to filter your experience in a manner that buffered your encounters with utterly different symbols as it attempted to mitigate the onset of disorientation or uncertainty. Perhaps the American tourist walking through Tokyo would *feel* more connected to the city if all the street signs changed to English before him. But how much of the translated city would still be Tokyo? Digitally scrubbing the place of all Japanese characters written on shop doors and building facades would surely limit the tourist's perception of the local culture. It would, reminiscent of Weiser's objection to interface agents at MIT, effectively outsource crucial acts of sense making, crucial aspects of tacit engagement. The tourist's sense of connection forged from sights of a familiar language might impede a more authentic connection attainable only through experiencing the unknown and processing the stark differences. A "world which is always familiar"—made up of places rendered bespoke by technologies designed to help you feel more at home wherever you go—could furnish a life of easy, illusory connections at the expense of more transformative encounters.

For the time being, Weiser listened to Suchman's criticisms but still forged ahead. Her hypothesis about the desires driving ubicomp may have hit too close to home. Consensual location sharing, he maintained, gave workplace acquaintances a window into others' lives. New points of contact allowed for more opportunities to get to know someone. Badges enabled their wearers to share otherwise invisible sides of each other. At some level deeper than he would've aired, Weiser's enthusiasm for the active badge may have come from a more plutonic place, in the same region of his heart that wished he could wear a "Love Wanted" sign on his chest, decades ago in his college poetry. Perhaps it resonated as the first step toward something like the thought bubbles he dreamed up to unite strangers in one of his short stories. Like PARC's badge network, those thought bubbles had been premised on reciprocity, and they gave rise to kind acts and better friendships—or at least they did in Weiser's fiction.

Weiser had also harkened back, in an essay he finished in January 1994, to the ideals exemplified in Polanyi's white cane that hooked him at age sixteen. There was still something fundamentally noble to him about any tool that made your surroundings more tacitly knowable. Loading his readers with metaphors and models for grasping ubicomp's basic rationale, Weiser drew upon his early influences when he wrote, "A good tool is an invisible tool. . . . Eyeglasses are a good tool—you look at the world, not the eyeglasses. The blind man tapping the cane feels the street, not the cane."[60] Theimer's hypothetical American walking around Tokyo could not perceive the meaning of the words describing everything around him. Was it such a transgression to offer him the linguistic equivalent of a white cane? Technologies that altered one's situated experiences—as well as, in some cases, the physical setting itself—certainly required justification. But the Responsive Environments projects and the debates about badges made it clear to Weiser that justifying ubicomp to the public would be as difficult and demanding as clearing its technical hurdles inside the lab.

Later that year, he demonstrated PARC's badges and sensors to a visiting journalist who then declared in *Smithsonian* magazine that "Mark Weiser might rearrange society as thoroughly as Thomas Edison did when he electrified our cities."[61] But in the months that followed that article's publication, Weiser hadn't even managed to make the rearrangements needed to grow the badge network throughout PARC. Brown stopped wearing his active badge after Suchman confronted him about it.[62] It had been easy to shrug off the worries of newspaper skeptics who manufactured their panic about badges from observations hashed together from a single afternoon. Criticisms from those on the inside were more serious. Weiser, Want, and their collaborators had pushed for an aggressive rollout in hopes of creating a tighter-knit community, and a part of that community pushed back with equal force. While the previous year's ubicomp retreat had been the last straw for Suchman and her fellow anthropologists, it was fully sinking in now for Weiser that the computer scientists and the social scientists would never become the big, harmonious, interdisciplinary family he fancied they might when he moved with his own family to Palo Alto.

His family, now, wasn't what it had once been either. Weiser had packed his suitcase almost weekly throughout the summer and fall of 1994. Organizers hosting technology conferences in Sweden, Poland, Switzerland, and nearly a dozen American cities had invited him to give keynote speeches about ubicomp. The titles of his talks varied, but he filled them with recycled content. The nomadic routine levied a draining personal toll that had also weakened ties back at home. And despite being always away on business—or perhaps because of it—Weiser seemed to need more time alone.[63] Weiser and Reich separated. Soon he was stuffing things into his car, moving out of the house he shared with her and their two daughters (though he would still see Nicole and Corinne often), and driving across town to live in a smaller place.

At PARC, he submitted plans for some research to throw himself into over the coming months. Six years serving as a lab manager had granted him paid sabbatical time to work solo for a while. The main items on his agenda included writing a book (publishers had been asking him for one), starting a business (no genuine ubicomp product had made a splash in the market yet), and creating ubi-artwork (why not?).[64] Newly relocated in Palo Alto and traveling constantly still, in December 1994 Weiser began his sabbatical from PARC, which he would wind up extending into an indefinite leave of absence.

7

Tacit Inc.

DURING THE MID-1990S, as millions tuned into popular sitcoms like *Seinfeld* and *The Fresh Prince of Bel-Air*, Lucy Suchman's research team gathered around footage of a lawyer and his filing cabinet. This lawyer had a reputation for being far and away the most organized person at his firm. Other attorneys popped into his office constantly to peruse the collection of several thousand legal forms and boilerplate documents, hoping to find one that might offer them a head start on whatever contract they had been assigned to create on their client's behalf. The firm had a motto—"If at all possible, avoid drafting anything from scratch"—and this filing cabinet was their best defense against the blank page.[1] Watching lawyers hadn't been Suchman's first choice, but she was eager to partner with a worksite and low on takers. She had hoped to land a labor union—the San Jose branch of the Service Employees International Union—but the urgency compelling the union's advocacy efforts left them too busy to take on a research project.[2] Anyhow, the process was what mattered most to Suchman and her team. They were piloting a novel approach to innovation that had not been tried in Silicon Valley.

The lawyers, with their massive filing cabinets and their constant need to locate the right file quickly, dealt with search-and-retrieval challenges faced in offices of all sorts. The Work Practice and Technology researchers were intentionally not regarding this law office as a stand-in for any law office, let alone any workplace that housed large collections of documents. The ethnographers, as they had done at San Jose International Airport, began with a meticulous study of each step in the workflow that brought a task to completion inside the law firm, from an attorney's first client meetings to the gathering and sorting of paperwork by the firm's document coders. But ethnography accounted for only half of the equation this time. Their observations around the law office would directly inform the development of an electronic repository for the firm's files, complete with a custom-made search engine to support rapid document retrieval. The team would be judging the success of whatever system they built *not* by how widely it stood to be used, adopted, and sold, but rather by how well it aided the legal work that got done within the confines of this singular building. The best way to do this, Suchman's team believed, was to cocreate the system with the lawyers and their staff from the ground up.

Suchman and her collaborators, Jeanette Blomberg and Randy Trigg, had drawn much inspiration from similar projects happening in Norway, Denmark, and Sweden. While American companies dominated the global computing market, the "Scandinavian model" developed better ways to involve office workers, service employees, and nontechnical professionals in the creation of hardware and software aimed at their jobs. The cooperative design community, which gathered annually for small international conferences at the fringes of the electronics industry, was still buzzing about a project launched in the 1980s by Pelle Ehn and Morten Kyng called UTOPIA. They had partnered with the Nordic Graphic Union to involve newspaper staff in the creation of a desktop publishing application designed to suit their particular needs and preferences. The newspaper workers were much more than a focus group—their union funded them to spend half their working hours over several years meeting with Ehn and Kyng to discuss every detail that went into formatting a newspaper. The workers and the technologists together visited labs in the United States (including PARC) to look at cutting-edge prototypes while they exchanged impressions about which new features seemed most relevant to their jobs. This relationship deepened as a pilot application was installed on computers in Swedish newsrooms and tweaked to the staff's liking.

Such a prolonged commitment to honoring workers' interest throughout a product's evolution went miles beyond any efforts Suchman had witnessed closer to home. Histories of American business chronicled the tactics of everyman executives and store managers who eyed shoppers as they browsed shelves of displayed clothing, kitchenware, and cleaning solutions.[3] Effective as it sometimes proved, snooping on customers was a method that still kept the customer far removed from important product decisions, as did later marketing research tactics like mass surveys and statistical analyses of sales data. The cooperative design project Suchman's team had taken on at the law firm set out to show that something like UTOPIA might work here, too.

Right around this time, beginning in the latter half of 1994, a chain of seemingly marginal developments on the World Wide Web drove a sudden uptick in online traffic. Few computer scientists around Silicon Valley gave it much attention at first. Top engineers at industry giants like Microsoft and Silicon Graphics were still rushing to build their interactive television sets, which every major CEO still believed would be the most popular vehicle for what they were calling "the information superhighway." Xerox PARC had no such initiative in the works, but that did not stop them from underestimating the World Wide Web's potential. Plenty of other projects did. In the area of hypertext systems alone, PARC scientists had invented several working systems that made Tim Berners-Lee's web seem amateur by comparison. The most revered of these in-house networks was called NoteCards. In addition to being more capable and sophisticated than that application running out of CERN, NoteCards predated the World Wide Web by roughly six years. Initially developed in 1984 by Frank Halasz and Thomas Moran, with help from Randy Trigg and Dan Russell, NoteCards allowed its users to group multiple texts—"notecards"—together within a single window. NoteCards was meant to hold smaller chunks of digital content so that, unlike with a web page, readers could click on a link to open a new notecard without leaving behind the initial notecard. Rather than have users jump from one page to the next, NoteCards kept both linked texts in view. Online files not only were connected but also were made to be viewed side by side in a way that showcased their relationships. "It could have been the web," said Russell. "Its big mistake was that it was a closed system; the web was open."[4] On account of the web's openness, programmers and hobbyists around the world could readily contribute to its development.

Up till that point, however, the web's openness had amounted to little. In early 1993, for instance, only a few dozen servers around the world supported access to it. Berners-Lee's name was largely unknown outside academic circles and a fledgling community of some two hundred technologists on the "WWW-Talk" Usenet group who chatted with him about their experiences tinkering with his software. Using the web's native markup language, HTML, anyone within this network could create their own website, though few more than fifty such sites had been created. Most were merely an alphabetized list of folders and files that could be scrolled through and opened one by one; many of the files were still scientific papers that researchers had posted for a smattering of peers. Images remained rare, except for occasional diagrams meant to accompany the uploaded transcripts of scholarly presentations. Berners-Lee was pleased with this progress, and he continued to view the web as a platform to connect the global scientific community. With an eye toward growing it in that particular direction, he issued an open call inviting anyone at the labs and universities within the web's orbit to improve upon the original web browser he had built three years earlier in 1990. His browser, it turned out, had been the main thing holding back the World Wide Web from spreading faster than any piece of software in history. The development of a better, more user-friendly browser hacked together by a few students in a basement at the University of Illinois would soon enlarge the scope of Berners-Lee's web far beyond anyone's expectations.

One of the first PARC researchers to suspect the web's eventual magnitude was Eric Saund. His specialty was computer vision, and he hadn't been interested in online developments like the University of Minnesota's Gopher system or the Whole Earth 'Lectronic Link, a text-based virtual community that counted many Bay Area technologists among its users. But Saund was struck by a new trend he noticed while sitting in the audience at a user-interface conference in November 1994, three weeks after a startup called Netscape had released its first web browser. Either at the beginning or end of their presentations, well over half the speakers displayed the address for their personal website. None of Saund's esteemed colleagues at PARC had a website, yet here were all these lesser-known interface designers, each belonging to a different lab from across the world, all pointing to their web address as if it were a natural extension of their identity. "I came back [to PARC] and immediately learned HTML, then told everyone else that we needed to do this," he recalled.[5] A year and a half passed without any reaction. It wasn't until 1996 that PARC held an internal workshop to walk its scientists through the process of creating a web page for themselves. From December 1994 to June 1996, the total number of websites would skyrocket from just 10,222 to more than 252,000.[6] "Xerox PARC stood on the sidelines," said Saund, "like we had nothing to do with the future of computing," even though their hallways were lined with the world's foremost experts in computer networking, toiling away on hypertext systems that would hardly leave the building.[7]

The web was not supposed to be a massive, all-purpose platform. Akin to early online services like AOL and Prodigy, the web's reliance on dial-up connections and low-bandwidth cables seemed to demand more patience than one could expect from the average computer owner. When Jim Clark resigned from Silicon Graphics to start a new company in 1994, he had intended to assemble a team of engineers to build software for the interactive television sets that several major tech firms were developing. Only after he met twenty-two-year-old Marc Andreessen, his would-be Netscape cofounder, did Clark consider the prospect of building something for the web. The author Michael Lewis has characterized Netscape's haphazard path to meteoric success as "one of the great unintentional head fakes in the history of technology."[8] At a coffee shop in Palo Alto, Andreessen clued Clark in to an exponential surge in web usage triggered by the Mosaic browser, whose creation Andreessen had spearheaded as a senior at the University of

Illinois the previous year. With Andreessen's vision and Clark's clout, the pair soon decided to bet Clark's money on the prospect that building a better version of Mosaic might accelerate the web's growth even further. If their improved web browser could excite a critical mass, then they would effectively forge their own shortcut to the information superhighway that much of the industry believed would arrive only on pricey new televisions—and not for another decade or so.

With their head start, Netscape wasted no time bringing their Navigator web browser into the world. Andreessen and Clark flew to Urbana, Illinois, where they hired six of Andreessen's programmer buddies within twenty-four hours. The next week, they were all coding away in an office above a Mexican restaurant in Mountain View, California. *How* they worked turned out to be just as influential as the product they made. Whereas the ethnographic immersion and participatory design tactics adopted by Suchman's team at PARC arguably represented the most thoughtful and considered edge on Silicon Valley's spectrum of R&D practices, the Netscape guys blazed a trail that led in the opposite direction. Their browser was hastily designed for anyone and everyone—whoever might happen to hear about it from an online message board, an email listserv, or chat room. Summarizing his start-up's mentality in retrospect, Clark wrote: "You conceived of it in your head, produced it in a computer, and tossed it up for grabs on the Net."[9] The young coders threw together the program's basic features—a home button, a back arrow, and the like—and they would later field emails from the first wave of users, sifting through the feedback for issues to resolve and new functionality to add on a rolling basis.

The Netscape developers' cardinal virtues were speed and responsiveness; they prioritized tending to their software's shortcomings one bug at a time as needed, instead of spending years crafting the technology, all while trying to better understand the needs of their intended customers. If something was wrong, they figured their users would message them about it. No news was good news. Any relationship more lasting than this seemed incompatible with the scale they were pursuing. Aside from constantly growing the user base at a dizzying pace, there was no grand plan for where Netscape would take the web, or where the web might take society. They were developing Navigator the same way they would've built a vast multiplayer video game. One of the founding programmers recalled, "We really didn't have any lives outside of the office so of course we're going to be at the office all the time!"[10] Andreessen's friends-turned-employees intensified their college work habits with a gold-rush fervor: all-nighters and binge coding sessions, fueled by caffeine and alcohol, that stretched over twenty to thirty hours bookended by naps under their cubicles or breaks playing foosball, air hockey, and Doom. All the while, Suchman's team diligently interviewed attorneys and legal aids at the law firm nearby in a daily effort to learn everything about the place so they might build a custom network that was precisely optimized for that single office. At the law firm, you had a team of renowned PhDs devoting several years to crafting a system for dozens of people. Above the Mexican restaurant, the Netscape guys banded together on a six-month coding sprint resulting in a globally adopted platform, without casting more than a passing glance at the wider world during their sleep-deprived commutes to and from their cubicles.

Local journalists published tales of their visits to Netscape headquarters, playing up the unusual office culture and rattling off statistics attesting to the company's emerging dominance. According to the industry veteran Brian McCullough, "It was this new paradigm for product development, more than anything else, that was Netscape's first contribution to the modern idea of 'a startup.'"[11] Having captured a majority of the web browsing market within six months of launching Navigator, Clark and the venture capitalist who backed him pushed to take the company public in August 1995. Netscape's stock jumped from $12 on the day of its IPO up to $140 just three months later, becoming one of the most lucrative in US history. Their triumph on Wall Street reverberated across Silicon Valley. Venture capitalists changed the way they assessed prospective companies. "No longer did you need to show profits; you needed to show rapid growth," wrote Michael Lewis.[12] Engineers with an idea left their salaried positions in hopes of making millions after a year of putting in hundred-hour weeks at a startup. The dot-com era had begun, and with it, the value of basic research labs—with their lofty ideals, academic leanings, financial agnosticism, and lengthy project timelines—was being called into question by the corporations footing the bill.

Another of Bill Gates's memos leaked from Microsoft in May 1995 and spread to inboxes across the tech industry, just as Gates's memo endorsing Weiser's vision had nearly four years earlier. In this latest issuance, titled "The Internet Tidal Wave," no trace of ubiquitous computing remained in the strategic priorities Gates hammered out. Whereas the PC market's momentary dip during the early '90s had channeled investments in mobile devices, the rise of the web, buoyed by Netscape's browser, lifted PC sales to new heights.

Microsoft's founder acknowledged that he was somewhat late in foreseeing the internet's revolutionary impact. Citing two observations he surmised from the web's recent growth, Gates declared: "The Internet is the most important single development to come along since the IBM PC. . . . It is even more important than the arrival of the graphical user interface."[13] The cost of accessing the internet, he first noted, was determined by the size and speed of your connection and "not by how much you actually use your connection." Anyone who could afford a connection faced no additional economic barriers once online. Unlike the telephone, you could use the internet all day without feeling a financial pressure to log off. Gates's second aha moment came from discerning "the positive feedback loop" just then taking shape on the web. "The more users it gets, the more content it gets," he wrote, "and the more content it gets, the more users it gets."[14] Indeed, this axiom about content and users—a digital-age rendition of what economists have long called "network effects"— anticipated the rise and reign of the web's top sites, from Yahoo to YouTube to Facebook. Gates ventured that the web stood to alter every other piece of software that had until then defined mainstream computing. "I want every product plan to try and go overboard on Internet features," he told his managers.[15]

The memo's theme resounded across all twelve of its pages: the internet in the form of the web was the next big thing—perhaps the biggest yet—and that made PCs more essential than ever, since desktops and laptops were the only consumer electronics that could support a robust online experience.

As Weiser scrolled through Gates's memo on his screen, he must've contemplated going with the flow. He was dead set on heading a startup that summer to bring a true ubicomp device to the mass market. But his business plan at that point amounted to little more than scribbles in his notebook. There was still time to pivot before pitching investors, and Weiser was no stranger to the web. His reputation as the visionary behind ubicomp overshadowed his involvement at the leading edge of early internet culture. Since the early '80s, his home computers were among the few household machines that were online then. In Maryland, he had been a mainstay among the first wave of contributors on Usenet's group message boards, and then he became an eager adopter of PARC's many in-house, experimental online communication platforms. One journalist reported that Weiser normally had eight video chat rooms open on his desktop at work.

More recently, the nation's prominent newspapers had been covering the digital exploits of his obscure Silicon Valley punk band, Severe Tire Damage, or STD—a name they had chosen largely for the sexual pun. Composed of four computer scientists in their forties, with Weiser on the drums, STD became the first musicians to broadcast a live show over the internet back in June 1993. The video feed they had rigged up gave the small audience of techies who had tuned in a real-time, highly pixelated glimpse of STD rocking out before an even smaller crowd of researchers gathered on a Xerox PARC patio. The event had a technical component, of course: it marked the first musical demonstration of an audiovisual broadcasting application called Multicast Backbone, or "MBone," which Steve Deering had been refining at PARC's Computer Science Lab under Weiser's leadership. While Weiser remained focused on ubiquitous computing in his own research, he had also been shepherding video applications for remote collaboration and livestreaming developed by his colleagues in the lab. MBone emerged as the most advanced of these platforms, and a few music-industry executives who learned about it believed it could be an innovative tool for promoting their biggest bands. STD lurched into global notoriety the following year when the Rolling Stones were getting ready to be "the first major rock band" to livestream a concert online, courtesy of MBone.[16] As fans around the world logged on early to see their favorite band walk on stage at the Cotton Bowl in Dallas, they saw instead the men of STD—standing with their instruments in some dim garage. Weiser's band had taken the liberty to hop on the Stones' video channel and play a few songs in the minutes leading up to showtime. They announced themselves to viewers as the Stones' opening act, much to the viewers' confusion. While the stunt garnered mixed reviews (a *New York Times* reporter quipped that the historic moment "was tarnished by a little-known rock band called Severe Tire Damage"), MBone became as much a part of the story as the music, and Weiser was quoted alongside Mick Jagger.[17] The Stones themselves seemed to have been amused; their spokesmen called STD's surprise performance "a good reminder of the democratic nature of the Internet."[18]

But Weiser's trailblazing forays online were, like his drumming, just a quick hit of fun. Never mind that MBone, with the capacity to broadcast live video online, seemed like the most promising solution to what Gates and others were dubbing one of the web's most wicked problems. Neither the lucrative challenges involved with streaming audiovisual content nor the new social frontiers of untethered virtual communities held Weiser's focus.

Weiser remained skeptical about the techno-optimism being heaped upon the web by throngs of enthusiasts who hailed Netscape as an opening to a better, other world. John Perry Barlow, the former Grateful Dead lyricist–turned–"cyberlibertarian activist," would soon author his "Declaration of the Independence of Cyberspace" from a plush hotel in Devos, Switzerland, wherein he informed all the world's governments that their "legal concepts of property, expression, identity, movement, and context do not apply to us."[19] (In hindsight, Barlow's "us" came to describe executives at Google and Facebook much more accurately—their general tendency to dodge taxes, exploit user data, and defy copyright protections—than the disembodied, anonymous Net dwellers Barlow had in mind.) Weiser dismissed prospects of a digital transcendence when *Fortune* magazine asked him in March 1995 to opine on the web's future, and the future proved him right a decade later. "As more and more business is conducted online," he said, "it will become more of a real place, and real-life expectations will take over."[20] Weiser had been sampling virtual worlds long enough to know that they would always leave him wanting. Whatever content a desktop served your way, you were still the person who was sitting there at the desk, in a society upon a planet, where life's sharpest meanings solidified on material grounds. After the venture capitalists won big by backing Netscape, technologists across the valley scrambled to populate it with companies that slapped ".com" onto pets, cars, birds, and babies. Just about any physical entity could inspire a website that might attract speculative investment. As Weiser plotted a direction for his startup, he had already decided on an opening statement with which his first presentation slide would confront potential investors: "This is not another Internet deal."[21]

Weiser backed his way into the formation of his company. He started by pondering a name and jotting some catchphrases to suggest the firm's essence. The first one he wrote down appeared too perfect to bother listing any more: *Tacit Inc.* The name implied a promise to himself that, with this venture, he would return full circle to his adolescent inklings of what technology ought to be back when he sat with Polanyi's book in silent wonder. Holding close to *The Tacit Dimension*'s reverence for embodiment and intuition, Tacit Inc. would specialize in devices that were, above all else, subtle. The true products were to be the newfound sensations of mental clarity, continuity, and flow their customers would experience once they integrated Tacit's humble wares into their daily routines. "Tacitness," Weiser emphasized in his notebook, "is more a property of the person than the technology."[22] The mission statement boiled down to helping people situate themselves in a more optimized relationship with their surrounding environment to quietly benefit their well-being. Weiser's first pass at corporate slogans conjured up a strange parallel universe of what TV commercials and product packaging might be like if researchers and academics were in charge of advertising their inventions.

> Tacit . . . We make your unconscious smarter.
>
> Tacit . . . It's like having lots of computers, dedicated to your different functions.
>
> Tacit . . . Providing stability, the essence of life.
>
> Tacit . . . Automatic transmission for your PC.[23]

By commencing with these rhetorical tasks, Weiser strove to tease out the reality of promises he knew the hardware could not manifest in its current state. What was possible within PARC's wireless network became five or ten years beyond reach the moment you left the building. On the inside, the stuff of Weiser's imagination pervaded the space, infusing offices and hallways with extra layers of meaning that, for him, made everything feel more in tune, connected, and alive. He could speak to the future it pointed to during his talks to fellow researchers, but such gestures would carry no currency in the venture capitalists' boardrooms. Preparing to sell them on Tacit Inc. forced Weiser to whittle his vision down to something that could function in the world as it now was. Almost nobody outside PARC had a hundred computers per room, and the one or two they did have were PCs. Their ID badges didn't adjust the thermostat; their alarm clocks and their coffeepots didn't collaborate; the views from their bedroom windows didn't feature any digital traces of their kids or neighbors. The first Tacit device, whatever it was, would have just one other object to converse with: ubicomp's path to market led straight back to the desktop.

By September 1995, Weiser deemed a few ideas ripe enough to share with venture capitalists over meals at the expensively rustic diners and cafes of Sand Hill Road. Flirting about new enterprises over a casual lunch had fast become a standard valley courtship ritual of this Netscape era. Weiser met with four VCs in two weeks.[24] He shared the happy news that Xerox was prepared to give him $500,000 in seed money, and that the brilliant Roy Want had agreed to serve as Tacit's chief engineer once they secured the additional $2.5 million investment needed to initiate production.

Exactly what Tacit should produce first was still something of an open question. Weiser walked each VC through three concepts he and Want were envisioning: buttons, displays, and keychains. Each category in this would-be product line aimed to help people streamline their relationship with their PCs. The industry's collective push to make the desktop a software-packed, general-purpose machine had made navigating its array of functions a cumbersome chore for all but those fluent in keyboard shortcuts. Adding Tacit's buttons to any PC, Weiser told the VCs, gave users a direct line to the programs they used most. If an artist or child designated a button for their favorite drawing software, they could physically click it with their finger—bypassing the mouse and desktop icons—and the program would instantly open. Similarly, Tacit's display concept imagined a cheap, small, wireless screen that showed just one select chunk of online information, such as a stock price or one's email inbox. Both were meant to give users the ability to prioritize a particular component of their PC by distinguishing it physically from everything else on the screen. The interventions were minor, of course, but they stood to make electronic tools a little more ready-to-hand. Both software and documents would be rendered more immediate—more like a hammer lying on a table than a hammer stashed in a toolbox stored in a shed with all manner of stuff.

The VCs advised Weiser to shelve buttons and displays for the time being and make Tacit keychains his first priority. The keychain promised to solve a basic problem that was growing with the web. More workplaces than ever were stocking their offices with desktops, and many American households followed suit. The contents of one's work computer, however, did not sync with one's home computer. Unlike today, there was no cloud-based storage platform onto which users could simply drag and drop files for access elsewhere. There were floppy disks and there were "personal digital assistants," like the Apple Newton. Both mechanisms had to be carried around, and both entailed wasted time. After inserting the disk or plugging the PDA into your PCs, you selected the files you wanted to move and then waited, often minutes, for the transaction to end. Tacit's keychain—an unremarkable plastic fob that weighed less than any metal key fastened to it—would make the file transferring process instant, automatic, and wireless. It would be like having your own little digital cloud that floated with you to every PC. Once set up, your keychain's proximity to any one of your devices would initiate the exchange of your most recently updated files using infrared light. All but one of the VCs urged Weiser to send them his business plan.

What excited Weiser most about the keychain idea was, in hindsight, of marginal interest to the VCs. For Weiser, this automatic updating of one's digital information ultimately gave people the means to spend less time looking at their PCs. The payoff he celebrated was the bypassing of all the idle moments users had to sit through, wherein the PC required them to monitor a process that shouldn't require their direct supervision. Augmenting one's intelligence through software still meant yawning through mindless procedures at regular intervals. Minimizing the latter, starting with keychain, would help. Weiser hoped Tacit's early offerings might also prove the concept that ubicomp would gradually liberate people from having to adjust their lives to fit their PC's technical specifications in order to reap the benefits of computing. A keychain here and a display there would equip them to incorporate smaller devices that were optimized around a single task. Increasingly, though, the VCs were banking on people staring at their PCs for hours on end.

Tacit's projected product line only hinted at the ubicomp scenarios Weiser thought technologists would be able to deliver in ten years. Still, the prospect of creating something that had immediate influence held ample consolation. Aiming toward an imagined future for the past six years had left Weiser desperate for some commercial validation, mainstream customers, or really any sign that his long-term quest wasn't too far afield. He was, he confessed in his notebook, starting to lose his confidence. He was also aware that interest in ubicomp at PARC had been waning: "My colleagues don't respect my research," he wrote, perhaps thinking back to the active-badge experiments.[25] In his notebook, he staked everything on the fate of his startup. "I can correct all [this] by proving myself in a new domain," Weiser confided as he raced to finalize the business plan and send it to the VCs.[26]

Weiser's proposal led with an epigraph—a faintly scholastic gesture in which he cited the enlightened desire of *InfoWorld*'s editor in chief, Stewart Alsop, as evidence for Tacit's sophistication. Just weeks before Weiser submitted the plan to VCs, Alsop had published an open plea for technologists to invent something exactly like keychain. Alsop pined for a tiny yet rugged device that could "suck just the data I want—the appointments and contacts and notes I need with me all the time—out of my computer, whenever I am close by."[27] Even more than the keychain itself, Weiser emphasized, Tacit's underlying product was "the automatic update of information."[28] His proposal went on to enumerate the licensing opportunities to be had by selling this capability to other companies who wished to incorporate automatic updating in their next generation of PDAs, telephones, and PCs. Through royalty deals and retail sales, Weiser expected Tacit to make its first sales to a segment of roughly ten million "mobile professionals" in the US whose constant travel to sales appointments, client worksites, and trade conventions positioned them to be keychain's earliest adopters.

However, a different set of business metrics tied to Wall Street was changing the way VCs spent their money. This ascendent vocabulary—*hits*, *clicks*, *daily unique visitors*, *downloads*, *IPO timeline*—signified a path to quick global success that did not extend to offline hardware. On the heels of Netscape's launch, the search site Yahoo found itself basking in the internet's spotlight when Netscape "decided to make Yahoo the default link when a user clicked the DIRECTORY button on the top menu of the [Navigator] browser."[29] These were moments that could make or break a new venture in the web's viral ecosystem. Within weeks, Yahoo had transformed from a Stanford student project into something that more than a million eyeballs fell upon every day. Sequoia Capital, the VC group that backed Apple in 1977, gave Yahoo's two boy founders $1 million for one-fourth of the company-to-be in April 1995. In the midst of such deals, Weiser was seeking an initial investment for well over twice the Yahoo amount just to get Tacit up and running.

To help Weiser better understand the mobile professionals that he assumed would be Tacit's first buyers, PARC convened seven white-collar nomads to give their impressions about keychain and how it might aid their work. The focus-group facilitator asked them to list the tools they used to take notes at meetings, track expenses, and record new-client contact info. The more-senior men said they just called their secretaries, whom they had standing by at an office computer, whenever they needed to know something. Others struggled to keep a filing cabinet's worth of printouts in their briefcases, though the heft made traveling a pain and finding the right document a chore. Two of the group confessed they had bought "fancy" electronic organizers—Timex's new Datalink watch and the Apple Newton—which they no longer used, out of frustration. The watch could hardly store anything, while the Newton ranged from being "a hassle," "non-functional," "glitchy," and "worthless."[30]

Weiser must've been giggling in agreement as he listened over the intercom in the next room. The group's quips squared up perfectly with problems that keychain was born to solve. "At this point," the facilitator told the room, "I'd like to invite in someone who will explain to you . . . a concept." Weiser entered and, without any introduction beyond his name, proceeded to describe "a very small device" that wirelessly transferred files to and from any PC, automatically. The room gasped in approval.[31]

Thrilled by the group's initial enthusiasm, Weiser exited the room once his speaking role was done and resumed listening over the intercom to the group's subsequent deliberations. Weiser had mentioned, almost in passing, that keychain would have a display screen, just in case you needed to look up a phone number or address while on the go between your PCs. Anything more than that—such as reading an email or typing one—was a job best saved for your desktop or laptop. The focus group discovered that they disagreed, once they started chatting about it. Within minutes, they had turned on Weiser's concept altogether. They were aghast now at how little keychain's screen aspired to do. One of the younger guys said, "I don't like it—it's too small."[32] What good is a device, the others agreed, that can only show you phone numbers? They joked about having to scroll through hundreds of numbers to find the one you needed. "You can't enter anything into it," chimed another. "It seems more inconvenient than convenient."[33] As the group strayed further from Weiser's prior calculation of their own best interests, Weiser complained in his notebook: "They self-limit their understanding . . . impossible constraints!!!"[34] Whereas the group had disparaged their experiences with PDAs like Newton, now they were railing against keychain for not being a Newton. They lamented the absence of graphics and color. The question of what the screen couldn't do had led them to forget keychain's core function. The prospect of automatic, wireless updates that had elicited awe when Weiser first mentioned it was suddenly deemed, upon collective reflection, "not compelling."[35] Mobile professionals were probably the wrong market to chase. While their jobs clearly revolved around bringing updated information with them wherever they went, traveling with the keychain would be of little use unless they brought a laptop along, too. In any case, eavesdropping over the intercom was the first time Weiser had paused to listen to the people he had designed keychain for.

Weiser was already bracing for rejection when letters from the VCs hit his mailbox that winter. Whether the VCs feared consumer reaction on a par with Weiser's focus group or whether they were just too busy chasing dot-com startups, their letters to Weiser didn't say. Tacit Inc. suffered death by a dozen platitudes. One missive after the next thanked Weiser for the opportunity to review his plan and commended its "many positive features," then confessed with some regret that "it does not meet our current portfolio focus."[36] Weiser stuffed the rejections in his filing cabinet, not far from a folder that held the timeline charting Tacit's path to breathtaking success, which Weiser had typed up when the idea was new. The document outlined a soft launch around the Bay Area during the summer of 1996, followed by national advertising campaigns fueling holiday sales that winter. These projections all led to a milestone in the summer of 1999, when keychains would, Weiser imagined, be running on 5 percent of all the world's PCs.[37]

Not a single unit had made it to stores; none had even been manufactured. Keychain would remain a diagram on paper, a picture painted in conversation, another working prototype. On the heels of his failure to launch Tacit Inc., the next item on his sabbatical to-do list acquired a somber tone: "Write a book."[38]

Weiser's father had tried to write a book once. When Mark was eleven, the family followed David Weiser to New Haven for his yearlong sabbatical at Yale. The house they rented on Long Island Sound was about twenty miles across the water from Stony Brook, where they would end up a few years later. When David wasn't inching his way through his manuscript, the family went for walks on the beach and played Monopoly after dinner. David brought Mark to a couple Yale football games. These outings were exceptions and work was the norm, and this generally left the kids sitting around the television. Mark took to wandering around the neighborhood with a fishing pole in search of a spot to pass the afternoon. Most of David's diary entries that year showed him agonizing over his book-to-be, right until the day he suddenly gave up. A letter came from Rand McNally, the publisher he hoped to win over. "They didn't care for the book and were sending the manuscript back," David noted.[39] After all his proceeding entries, the struggle and aspirations logged over the preceding months, he shrugged at the rejection. "It's good to be out of the thing. What a fearful, miserable mistake it was."[40] Writing to himself about himself became David Weiser's main subject from that point on; detailed accounts of cigarettes and alcohol, coupled with reflections on hangovers, gradually became his central motif. Mark and his sisters had helped type up the manuscript and checked over the spelling to make sure it was all correct. There was little talk of the book after that.

Mark Weiser now toyed with the notion of writing a brief memoir about Tacit Inc. called *Lessons from a Silent Startup*. He scribbled out a table of contents.[41] The exercise grew all the more fraught when Nicholas Negroponte's new book started climbing bestseller lists; it was just a few years before that Weiser had joked about the shortcomings of Negroponte's butler on stage at MIT. Compiled largely from columns Negroponte had written for *Wired* magazine, the freshly printed 1996 paperback edition of *Being Digital* became the year's most fashionable accessory for Silicon Valley's laptop-touting set. *Being Digital* cast a series of stunning predictions about how the internet, multimedia networks, and interface agents would swiftly alter various industries and facets of society. Negroponte's hypotheses for the future of television and news, for instance, proved correct. TV would eventually morph, Negroponte foresaw, into "a random-access medium" driven by streaming content available on demand, while the delivery of daily news would break from the common mold of regional papers in favor of online services that aimed to curate an idiosyncratic selection of items filtered around each person's interests.[42] The emerging norms of the digital world would outperform traditional media and rightfully replace merely physical artifacts. Samuel C. Florman's review in the *New York Times* was not the only piece of commentary to deride Negroponte's tendency to "celebrate information while disparaging the material world."[43] If any thesis could be said to underlie the book's roving futurism, it was Negroponte's opening claim that bits of digital data were becoming more central to humanity than the atoms that gave shape to life off screen. "The change from atoms to bits," he insisted, as if the process were natural, "is irrevocable and unstoppable."[44]

Probably no other statement by any other person could have incensed Weiser more, though the rest of the book marshaled a parade of contenders: "We will socialize in digital neighborhoods in which physical space will be irrelevant"; "computers will be more like people"; "we will find that we are talking as much or more with machines than we are with humans."[45] Big names across the media and business worlds who had never heard of ubiquitous computing began singling out Negroponte as the visionary for twenty-first-century tech. The stakes of such a popularity contest loomed far beyond esteem.

To Weiser, the buzz around *Being Digital* gave lay readers a warped first impression about the purpose and value of so-called smart objects. It was as if the primary ideas that colored images of ubicomp still unfolding in Weiser's head were being painted over and filled in, gaudily and ungracefully, faster than he could keep up with. The guise of interface agents—symbolized by the butler metaphor Weiser despised—enveloped nearly all Negroponte's descriptions of post-PC interfaces that a person might wear, carry, or be surrounded by in the coming decades. Interface agents in your earrings or your cuff links, for example, would recognize your voice, listen for verbal cues, and likely speak back.[46] The promise of moving beyond the desktop, in Negroponte's rendition, was the promise to bestow the simulacrum of aristocratic and executive privileges onto the middle class. Still peppered with references to butlers, chauffeurs, and personal assistants (stuffed between mentions of cappuccino, Swiss villages, and Evian bottled water), Negroponte's sentences bandied the notion that having servants confined only to the rich was a sweeping historical injustice that couldn't be righted soon enough.

He hinted at scenarios wherein digital domestics would be eager and ready to serve, fulfill, and anticipate the average user's every desire. "The idea that twenty years from now you will be talking to a group of eight-inch-high holographic assistants walking across your desk is not farfetched," Negroponte assured.[47] It was largely the same pitch for "intimate computing" that Weiser had criticized on stage in that MIT auditorium back in 1992. Interface agents wedged themselves at the center of attention between you and the physical world, and they promised to stick with you—in your ear, by your side, or in your face—everywhere all the time. A harbinger of AI, they purported to handle a lot of the thinking for you, so long as they could learn from an ever-accumulating pool of data they gathered about you, on their own initiative, just as any butler worthy of Negroponte's appreciation would do.

A hurdle inherent to ubicomp had always been the lack of technological infrastructure to maintain connections between wireless devices on a meaningful scale. Now, as time's passage ticked closer to overcoming that, the more daunting impediments to Weiser's minimalist ideals were coming from smart technologists with eminent clout, who couldn't fathom the thought of leaving bells and whistles on the cutting-room floor. At Negroponte's Media Lab, like in most of Silicon Valley, smaller was often better, but less was never more.

As the tech world pressed forward, Weiser found himself thinking back to an Ethernet-connected string that once dangled from the ceiling in the hallway outside his office door at PARC. The string experiment had been conceived as a work of art, more or less, and to Weiser it was. Slowly, Weiser's fond memories of the string, and of the artist who designed it, became more interesting to him than any book he might coax himself to write about Tacit Inc.

8

The Dangling String

WHENEVER WEISER VISITED AN ART MUSEUM, he bypassed the paintings in search of sculptures. To his eye, objects crafted in three dimensions exhibited their creator's touch to a greater degree. He could walk around a statue to see it from all sides, and each angle brought new details into view. The larger ones contained elements—like the sling Michelangelo carved upon *David*'s left shoulder—that played with the surrounding light to form shadows upon themselves and the ground. Lingering beside a great sculpture opened his mind to generous new appraisals of physical objects. A mute, inert thing became like a poem in space when given enough thought. It could impart morals and values, set moods, and even cast subtle rhythms that people would instinctively pick up on together. Sometimes Weiser liked to stay late at PARC in the evening until it was nearly empty, when he would pace through the dim corridors, the meeting areas, and the silent auditorium just to imbibe the building's architectural character.[1]

One night, before he had gone on sabbatical, Weiser turned away from the emails on his office computer to study the noise humming outside his door. It was the whirling whip of a hollowed-out red audio cable—or, as he later described it, "an eight-foot piece of plastic spaghetti"—that had been installed in the hallway a few weeks earlier.[2] No one else was at their desk, yet the cord was flailing hysterically, and something was suspicious about that at such a late hour. Weiser notified the system administrator. The cable's erratic hum, it turned out, was in fact the online footsteps of a hacker who had broken into PARC's local area network. It made for a funny story because this plastic spaghetti contraption hadn't been designed to catch digital thieves. Initially, nobody really knew how to explain its purpose, and some doubted that it had one.

The red cord dangled from a tiny motor mounted in the ceiling, and this motor was connected to the Computer Science Lab's Ethernet. Whenever the network's traffic spiked—while, say, a couple researchers were uploading huge files or streaming video—the cord would twitch, spasm, and then spin in a rapid circular motion. Weiser began to hear in its movements a kind of ambient music.

There were familiar beats that repeated across certain phases of the workday. During busy times, a strong, electronic wind blew from the cord's centripetal acceleration. That distinct pitch offered each person working alone in their office an instant explanation as to why their computer had just slowed down—it was just that the network was congested; their machine was fine. As afternoon became evening, Weiser grew to anticipate the relaxing pitter-patter of the cord's faint, intermittent jazz. It was the sound of a couple researchers like him who hadn't yet gone home. Sitting at their separate desks staring at screens, each one's online activity made the red cord pulsate. It was a bit like the feeling of being in a band. "Mark actually really loved it, and loved becoming a virtuoso in how to interpret its strange language," said Natalie Jeremijenko, the artist who dreamt up the unusual device.[3]

Jeremijenko had named her creation *Live Wire*, but Weiser liked to call it *The Dangling String*. This was not Jeremijenko's first such exhibit. Before blazing into PARC, she had sent fire zipping through audiences at a popular arts and music festival she cofounded in Australia. She and her festival staff rolled ropes in gunpower, then wrapped them in cellophane; "every now and then," she said, "I'd drop a match on them and fire would sear through the crowds."[4] The festival grounds became her studio for unleashing all manner of "social sculptures" designed to jar the inebriated adolescent hordes. She was working on a PhD in neuroscience at the time—the first of four doctoral programs she ultimately entered—and from those initial artsy antics in Brisbane followed a museum's worth of discipline-bending experiments that, once Jeremijenko left PARC for academia, would come to occupy the walls and grounds of many galleries.

On her desk at PARC she kept a hand grenade. Those who asked about it were relieved to learn the grenade contained no explosives. It had been hollowed out, but it gave Jeremijenko occasion to talk about her mission to "demilitarize technology."[5] Too often, the knowledge and intelligence that computer systems housed were inherently concealed from plain view, let alone public access. To the extent that it further empowered a technocratic few over the many, computing could and did serve as a weapon of sorts, she insisted. That notion did not generally sit well with her computer scientist colleagues. Where Jeremijenko saw hardware and software that enabled hierarchical control at an unprecedented scale, they saw themselves as mavericks bringing the stuff of science fiction to the people, or at least to like-minded tinkerers. "The Bible of common reference at PARC was *Star Trek*," Jeremijenko recalled, "and frankly, I didn't really much like *Star Trek*."[6] She wished to design things that showcased—publicly and beautifully, in shared physical spaces—the *meanings* to be gleaned from computational data. Only then would digital information blossom into a truly democratic medium for exchanging and generating knowledge beyond the confines of institutional power, economic privilege, and technical prowess.

Most of researchers in the Computer Science Lab dismissed Jeremijenko's point of view; they remained convinced that personal computers were already the boon to society that she was striving for in her art. After she presented her work they would sit in utter silence, offering no thoughts or feedback. "I felt like I was talking to aliens," she said, "except for Mark [Weiser] and Rich Gold."[7] Jeremijenko's *Live Wire* project had only worsened her standing with the engineers who found her frivolous. Even the colleagues who helped her get *Live Wire* up and running, by reverse engineering the lab's Ethernet switchboard, made her promise never to mention their involvement to anyone, for fear they would be ridiculed by association. "I was a young kind of girly girl," said Jeremijenko, "and I was just looked at with suspicion by almost everybody [in the Computer Science Lab], except Mark. . . . As soon as Mark went on sabbatical, I was fired."[8]

Weiser felt adamant that Jeremijenko brought a certain aesthetic sophistication that PARC lacked. Their collaboration developed into a friendship when they soon recognized elements of themselves in each other. Both Jeremijenko and Weiser had learned computer programming on punch cards at an early age (Jeremijenko in second grade), and both came of age through what they described as an intellectual crisis, where their youthful reverence for scientism had been fractured. Through the cracks seeped art, philosophy, and social critique exposing the partiality of scientific frameworks. Neither was deterred from a career in science, but the breach in their faith left them feeling sometimes estranged from dogmatic colleagues. "There's something that forms between people who have been very immersed in the technical sciences, but who are also seriously trying to understand culture," said Jeremijenko. "There's a secret society of us."[9] The garage of her house near Stanford—stocked with throwaway gadgets and freebie parts like some Goodwill store for discarded tech—became an evening hangout for younger researchers who had been raised on tales of garage-based breakthroughs. Weiser stopped by often to resume the conversations he and Jeremijenko had started earlier in his office.

As the two of them fiddled with capacitors and wires, they batted around tough, basic questions: "What is the job of technology? What is it for?" And Jeremijenko pushed Weiser to think further about ubicomp's possible cultural payoffs: "What is this change? Why is it good? Why would we want to have more computers everywhere?"[10] He really hadn't given complete answers in his essays after all, and no one else would press him on it like that, so bluntly and profoundly, during any of his monthly keynote speeches around the world.

Just before joining PARC, while she was pursuing a PhD in the history and philosophy of science at the University of Melbourne, Jeremijenko conducted an ethnographic study looking at the ways neurosurgeons used the large-scale data-visualization program she had created the previous year. Her program aggregated CT scans and histological slides from thousands of congenitally deaf patients, in hopes that the interface would help the doctors gain greater insight into the auditory cortex, the temporal bone, and other areas of the brain related to hearing. She hypothesized that the more information she could squeeze in, the more powerfully her system would aid medical research. And so it was rather devastating to learn that while the hospital loved to show off her mesmerizing diagrams to donors and prospective students, the doctors found no use for them. "It looked cool, like something from a Steven Spielberg movie," she recalled, "but none of [the doctors] knew how to read it."[11] The mathematical complexity layered into the program ventured far afield from the statistical literacy they had cultivated in medical school. Jeremijenko's interface, though packed with relevant data, was simply incompatible with the neurosurgeons' working context. When she first came across Silicon Valley's rally cries—axioms like "Information will set us free!"—she regarded them with a skepticism painfully learned from that Melbourne hospital. Digital data, however big or small, was valuable only insofar as people were able to grasp it amid the variables and vectors of their hectic lives. Having information present was just one part of the equation, and figuring out the rest was absolutely essential. "The lie of the Information Age," Jeremijenko would say, "is that more information is better."[12] What was needed most, she and Weiser began to believe in her garage, was a new design philosophy founded upon a different theory of mind.

Lately, with his sabbatical almost done and Jeremijenko gone, Weiser found himself fixating on the memory of *The Dangling String*, or *Live Wire* (they each still had different names for it). He was trying to extrapolate a larger idea from its little movements. The red cord's literal connection to PARC's Ethernet raised questions about other connections between his ubicomp vision and the internet's present state. Weiser's ill-fated brush with the venture capitalists and their lust for promising dot-coms had forced him to reckon with the web's imminent reign over all things digital. In the wake of Netscape's Navigator browser and more recently Microsoft's Internet Explorer, the global file-sharing repository hacked together by Berners-Lee quickly mushroomed into an infotainment carnival. To dial up the Net was to enter an unending maze of links and surprises: Anonymous chat rooms. GeoCities homepages. Grainy photos of nude celebrities. Dancing-baby GIFs. The musings of previously unpublished voices opining about unknown topics, all in assorted fonts and color pairings that flouted every rule of typography. Weiser and Brown summed up the web's clamor in a single image: "Late at night, around 6 A.M. while falling asleep after twenty hours at the keyboard, the sensitive technologist can sometimes hear those thirty-five million web pages, three hundred thousand hosts, and ninety million users shouting 'pay attention to me!'"[13] All inroads to cyberspace fed into the same feeling.

Riffing on the information superhighway metaphor, he and Jeremijenko had made a game of talking seriously about the informational character of various Bay Area locales. She would point out scenes in the valley below from the big windows of Weiser's corner office, and a thought experiment ensued. He remembered their comparative analysis of Highway 101 and the Junipero Serra, the two freeways that tied Palo Alto to San Francisco. Like the web, the 101 felt as if it had set off a big bang of colliding messages that now lined the ten-lane road. Billboards showing hamburgers and others displaying jewelry perched above buildings and towered up through the trees, one on top of the other. Add to that the traffic and glassy headquarters competing for one's gaze, navigating that stretch of the 101 invariably scattered the brain. Only three miles to the west, the Junipero Serra differed pleasantly. Signage was kept to a useful minimum that made it easy to consult when needed and easy to tune out when not. A slow reel of ponds, open spaces, and woodlands hugged its bends. Both freeways had been built using the same raw materials. If asphalt could be laid in the service of such divergent paths, surely the internet, too, might prove amenable to an alternative configuration that ran parallel to the World Wide Web. The spirit of those conversations had stuck with each of them as they returned to their work—Jeremijenko to the unsuspecting items in her lab fated to become art, and Weiser to his essay drafts and presentation slides.

Of all the eccentric things Jeremijenko designed at PARC, Weiser realized, *The Dangling String* stood out so merrily in his head because it bore a family resemblance to the pantheon of technical objects that had since college served as archetypes on his quest to imagine better tools. At first glance, the project might have seemed incongruous with the goals of ubicomp, since it drew attention to a technical apparatus. By giving dynamic expression to Ethernet traffic, the twitching wire didn't exactly disappear. On the other hand, as Weiser came to know it, the project showed how computation could take the form of a kinetic sculpture whose movements subtly conveyed the meaning of digital data, which people could understand instantly, without needing to look at a screen. *The Dangling String* converted the network's activity into a more natural sensory experience. What was cryptic suddenly became tacit. In Jeremijenko's lively red cord, Weiser saw an internet-age equivalent to Polanyi's white cane. Both devices made you aware of things in your surroundings that were difficult to perceive otherwise, and it was *the way they did it* that Weiser found so compelling. They indicated a method that could be generalized, and he had been searching for one. Weiser had at the time been obsessing over a question that had grown urgent amid the web's rise. Gold remembers him asking: "Can we find a method of information presentation that doesn't cause [information overload, frustration, and anxiety]?"[14]

Gleaning knowledge from *The Dangling String* was an "encalming" experience, Weiser and Brown would say, especially compared to surfing the web.[15] Jeremijenko's wire translated obscure flows of digital data into subtle audiovisual events that could be perceived with minimal effort. The sights and sounds it generated were, like the white cane's vibrations, more akin to wind in the trees than, say, a pop-up window or a dinging, flashing notification. Live indications of the lab's Ethernet traffic became part of the office environment and no longer a digital tidbit to repeatedly open up, click on, and close out on one's desktop.

The Dangling String demonstrated, more provocatively than anything Weiser had built, that digital information didn't need to be conveyed through virtual windows. Much would, of course; but some of it was worth weaving into other materials. Practically any object might serve as a fitting avenue for people to glean streams of data whose basic implications were clearly and fluidly put on display in relevant settings. To Weiser, Jeremijenko's art contained the kernel of a breakthrough in human-computer interaction: online data could be made to move like weather through our built environments, gently manifesting the information pertinent to each place.

Just as crucial for sustaining the calm awareness it engendered, *The Dangling String* did not outsource the task of sense making to artificial intelligence. This particular strand of calmness associated with Weiser's favorite technologies was not one of meditative withdrawal. It was, rather, a flow state, a sense of easy engagement, that accrued from the heightened sensitivity to key variables that a good tool afforded one's body and mind. AI overwhelmed that connection in favor of intricate black-box calculations; the system took primary responsibility for perceiving a situation, interpreting the meaning of it, and dictating a proper course of action to its human bystander. If the promise of AI lay in its speed and precision, its underside was how it rendered people. Someone who is, via AI, relieved of the burden to stay attuned to the unfolding present is effectively robbed of their agency. The AI user—like the "driver" in an autonomous vehicle—is positioned to be reactive, deferring to algorithms and interface agents rather than grappling with the world. Such AI systems discount the value of intuition, and they do not prioritize boosting one's tacit awareness. *The Dangling String* was about doubling down on both those fronts, and Weiser was determined to build on its methods and bring them to more objects.

In the summer of 1996, Weiser began negotiating with Brown about a new position at PARC. Weiser had already decided shortly after leaving that he would not resume managing the Computer Science Lab. He wished to focus entirely on taking ubicomp to new heights. The center as a whole, meanwhile, had been shaken up by changes a long time coming. The Xerox slush funds that generally allowed PARC to entice hotshot researchers, fly them around the world first class, and let them all work more or less on whatever projects they wished—that money had leveled off considerably since the heyday of Xerox's stronghold on the copier market.[16] Less was left over now for Palo Alto to play with, even as their R&D efforts still generated a sufficient payoff to justify the arrangement. The pressure to become more entrepreneurial, coming from Xerox headquarters and from the startup zeitgeist in the valley, was squeezing the spirit of basic research and long-term investment that had always distinguished PARC from the other corporate labs nearby.

Brown had turned Weiser on to motorcycling a while back, and as a pair they raced after work along back roads as the sky pinkened against the rolling amber farmland. It offered a release from the mounting tensions at the office, as well as another kind of experience to philosophize about. "There is a tremendous sense of calm," Brown said, "in streaming through the hillside at one hundred miles an hour."[17] Weiser agreed. As he had with Jeremijenko, he reflected with Brown on their daily routine with a phenomenological intensity. There was a local road they loved called Skyline Boulevard that careened through a misty forest on a ridge within sight of the Pacific Ocean. The twisting descent harbored an array of potential hazards, such as crossing wildlife or puddles that slip up your tire on a sharp turn. The speeding motorcyclist had to perceive, process, and react to all these surprises on a split-second basis. Weiser and Brown would compare notes after the ride. As Brown recalled, "We'd come back and we'd say, 'What is that? Somehow that never overwhelmed me. I never felt information overload.'"[18] They marveled over how this sense of calm persisted, despite being faced with far greater information on the bike at any given moment than even an hour's worth of web browsing. "That's where we also began to realize the difference between attending versus attuning," said Brown.[19] Surfing the Net was purely a matter of *attending* to the parade of content that you ushered onto your screen. You couldn't productively *attune* to websites out of the corner of your eye while you were attending to something else—and trying to do so eventually led to information overload. Descending Skyline Boulevard, on the other hand, entailed much more stimuli than one could ever hope to scrutinize closely at that velocity. And yet, as long as you were paying attention, the sights and sounds that were most consequential to your safety naturally jumped out at you. You attended directly to a small fraction of what was going on, but somehow your senses were able to *attune* to the rest.

Between the motorcycle rides and *The Dangling String*, their conversations acquired a momentum that compelled them to the blank page. They came up with a term—*calm technology*—then batted it back and forth to tease out its meanings. Whereas the notion of ubicomp introduced in that 1991 *Scientific American* article had gotten technologists excited about the future of post-PC devices, the prospect of developing this newly coined phrase into a full-blown theory seemed like an urgent follow-up. It was an idea that might cut right to the heart of the big questions that ubicomp raised and left unresolved. What would be the most desirable version of a world filled with digitally enhanced products of all shapes and sizes? Could environments be augmented and automated in ways that improve lives without creating a dystopian surveillance state in the process? These dilemmas no longer felt hypothetical, even for researchers outside of PARC, as several R&D labs were now making headway on the technical solutions needed to computerize everyday objects and connect them wirelessly to one another.

On August 14, Weiser's smiling face was printed out in press releases and newspaper articles announcing his appointment as PARC's chief technologist, a role freshly carved out for him by Brown. The job description, broad and vague, was essentially Weiser's to make up as he went along. His overarching duty was to represent the best of PARC's research to the wider world. When the press asked for comments about his plans, he alluded to the internet. "The Internet revolution has barely started," Weiser told reporters. "It won't be done until everything is on the Web."[20]

Weiser returned to the blue sky of basic research to find new agendas dawning. While the VCs had been stewing over Weiser's business plan for Tacit Inc., young academics and wide-eyed students were pulling all-nighters, pursuing the dream of an internet-studded world. Technology conference–goers no longer fell silent with amazement at PARC's wireless network of connected office objects. More technically audacious demonstrations of mobile and wearable computing were now happening at MIT. A recent cohort of rising-star graduates from New England's top universities had chosen to stay east. Thanks to an innovative funding model, MIT's Media Lab thrived in spite of the declining government and corporate support that was forcing other research centers into crisis mode. On the coattails of Negroponte's guru status, the Media Lab sold membership packages at six or seven figures a pop to giant companies eager to buy an early glimpse of the professors' inventions. Paying the annual dues granted these one hundred–some sponsors—firms including LEGO, Nike, Eastman Kodak, and AT&T—an invitation to exclusive project showcases and privileged access to the lab's considerable intellectual property. A reporter who visited the Media Lab during one of these members-only events likened the visiting executives to "children let loose in Willy Wonka's wondrous chocolate factory."[21] This mutually lucrative arrangement offered Negroponte's freshly courted hires that irresistible blend of security and flexibility, topped with instant prestige and the allure of evenings mingling with the Fortune 500. They decided Negroponte's ivory castle in Cambridge would be a better home than any in Palo Alto to build upon ideas they had read about in Weiser's essays.

At a 1995 gala celebrating the Media Lab's tenth anniversary, Negroponte had announced to an auditorium filled with his sponsors that the lab would be changing gears. The goal driving its first decade—to spearhead the merits of bits over atoms in hopes of propelling online multimedia into the mainstream—had culminated with the success of Negroponte's recent book. Almost as soon as *Being Digital* became a bestseller, he told the *New York Times*: "Our multimedia mission is over."[22] For its second decade, the Media Lab intended to shift much of its focus to projects that conjoined bits and atoms. The most heralded of these initiatives was named Things That Think.

Led by the professors Neil Gershenfeld, Michael Hawley, and Tod Machover, the Things That Think consortium struck up a collaborative alliance between forty corporate sponsors and several MIT research groups who were figuring out different ways to imbue everyday objects with digital connectivity. A physicist by training, Gershenfeld joined the Media Lab after collaborating with Machover on the design of a "smart musical instrument" for the renowned cellist Yo-Yo Ma. To Ma's cello they added sensors and a small antenna, which precisely tracked his fingertips as he pressed upon the bow. The cello itself gave off minimal sound; it functioned rather like a desktop keyboard. The electronic accessories measured Ma's playing and transmitted this information to a computer. The computer fed into speakers that emitted the audio. This elaborate feedback loop gave the musician more precise and more varied ways to watch his own performance.[23] Hawley also brought a musical mind to bear on the Things That Think agenda. He had returned to his alma mater after sharing a house with Steve Jobs and working at NeXT, while holding piano concerts on the side. Now a trio, Gershenfeld, Hawley, and Machover set about to engineer useful little symphonies out of the myriad interactions people routinely had with their stuff. They aspired to connect objects and harmonize the data that each generated to orchestrate a simpler yet enhanced way of life.

Gershenfeld believed that smart clothing—shirts and ties indiscernibly laced with microchips and sensors—could ultimately render portable computers obsolete. Clothes traveled effortlessly with the body; of all the things one could computerize, clothes offered a versatile platform for bringing computers into the fold of daily experience. Gershenfeld reasoned that shoes were the most fitting place to start—"feet bottoms make much more sense than laptops," he insisted.[24] One could embed a microchip in the sole without altering a shoe's look or feel; people constantly wore them, and the contact shoes made with the ground during each step offered a natural way to exchange information with other connected objects in the surrounding environment. With his student Tom Zimmerman, Gershenfeld discovered that smart shoes could also power a "personal area network" that allowed digital information to flow through one's body and be transmitted between people by touch. Businessmen testing their prototypes could exchange digital business cards simply by shaking hands.[25] From this initial demonstration, Hawley was quick to conceptualize a robust "Body Net" that would employ a full suite of wearable gadgetry to link one's bodily action with digital networks near and far. He imagined a lapel microphone, tucked under one's collar, that users might talk into in order to communicate with other connected objects nearby, as well as a smart belt buckle that could upload photos to the web when users pointed a smart camera at it.[26] Gershenfeld and Zimmerman liked to conjure up the scenario of a smart refrigerator filled with sensor-equipped grocery items: a milk carton might track its own contents and, when the milk was running out, it would notify the refrigerator, which would in turn notify its owner to purchase some more as he walked into the kitchen.[27] And, just in case the need to buy more milk slipped the owner's mind, the refrigerator could also be programmed to transmit its notifications to his usual grocery store, such that when he entered the store—when his computerized shoe stepped upon the store's computerized welcome mat—he would be reminded once again to get milk.

Other brilliant scientists around the Media Lab were aspiring to teach entire rooms how to think. Five "smart rooms" had been constructed recently by the Perceptual Computing Section, a group of more than fifty researchers led by Alex Pentland. What made each of these rooms smart, Pentland explained in a *Scientific American* article, was the room's capacity to monitor its inhabitants at all times, understand their every action, and "react intelligently to them."[28] An office of this ilk, Pentland suggested, might take steps to shield you from unwanted interruptions once it sensed you were on an important phone call. As soon as it recognized urgency in your manner and tone of voice, it might close your door and silence any incoming email notifications. Cultivating this brand of architectural omniscience involved a good deal of well-placed cameras and microphones, each always recording and relaying footage to a network of computers nearby. Pentland's group assigned these computers a distinct task: one analyzed the audio streaming in from the room, others made sense of gestures, and so on. The software they programmed made its calculations using a tactic called maximum likelihood analysis. This method relied on a database that contained models of facial expressions, spoken words, and other loci of human activity stored in the computer's memory. These models constituted a working vocabulary through which the software would seek to match the content of incoming footage captured from the room with the stored model that most closely approximated the live sound or gesture in question. Pentland's smart rooms, which borrowed from Pattie Maes's earlier trailblazing work on interface agents, had become quite accurate. The system could identify individual faces in a crowd of several hundred with 99 percent accuracy, and it correctly gauged expressions at a 98 percent rate.

Another of Pentland's projects brought this technology to cars. Collaborating with researchers from Nissan, Pentland's team helped create a program that monitored a driver's actions—all the movements of her hands, legs, and even eyes—in order to constantly anticipate the driver's next move. For instance, if the program detected the driver was about to change lanes, it would be primed to alert her of any vehicles in her blind spot. "Our smart room machines can answer a range of questions about their users," Pentland boasted, "including who they are and sometimes even what they want."[29] A former student of his was breaking new ground on this latter front: the Media Lab had retained Rosalind Picard to stay on as a professor to launch what she termed "affective computing"—a line of R&D that hoped to give computers "the ability to recognize emotions as well as a third-person observer."[30]

Weiser and Brown needed more examples to both stimulate and ground their ideas about calm technology. As they sat down to develop

MIT Media Lab staff testing a facial-recognition system for deployment in Alex Pentland's smart rooms. Photograph © Sam Ogden. Used by permission.

their motorcycle-ride conversations into a scientific paper, the headlining prototypes bubbling up from the Media Lab's new wave of researchers held some obvious appeal. The functionality exhibited in their celebrated demos handily exceeded PARC's ubicomp artifacts. Media Lab smart objects could do things that Weiser could only describe in thought experiments; Pentland's smart rooms commanded a degree of intelligence that hadn't even been on the radar of PARC's Responsive Environments project two years earlier. And, while MIT's avant-garde surpassed the technology generated by Weiser's lab, the Cambridge crew had adopted as their starting points many of the premises about computing that Weiser laid out in his articles from the early 1990s.

A simulation demonstrating a "smart car" program at Nissan's Cambridge Basic Research facility in Cambridge, Massachusetts, developed in partnership with Alex Pentland's team at MIT. Photograph © Sam Ogden. Used by permission.

There was so much of himself to recognize in their writing. They often began their essays and talks with Weiseresque jabs at desktops and laptops, that they were too isolating, too demanding, too severed from the fuller bloom of dynamic humanity. They shared the belief that a collective sense of information overload—which had spread fast with the web's exponential growth—presented a looming problem that would muddle online networks of all sorts until technologists developed more nuanced interfaces. Echoing Weiser's adage that "the most profound technologies are those that disappear," Gershenfeld everywhere championed the virtue of inconspicuous, invisible systems over and against the call for people to live pixelated lives.

Perhaps most appealing of all, it was clear to Weiser and Brown that these Media Lab projects desperately needed a notion like calm technology. For all their astonishing technical qualities, the disparate character of their applications made them like a series of experiments in search of a larger theory. Calm technology could provide a unifying rationale to guide these disparate experiments toward a common purpose. Just as the ubicomp concept had distilled a coherent vision from the fledgling "tiny computer" initiatives of 1991, this new concept might fuse researchers at PARC, MIT, and elsewhere around the notion that connecting a myriad of things to the internet could help people feel more attuned to the world around them.

A series of questions raised by Rich Gold had complicated the matter, as Gold's inquisition exposed a side of the Media Lab's aspirations—Pentland's smart rooms, in particular—that didn't fit so well with the whole "calm" ideal. Weiser and Brown couldn't unsee the hand-drawn images of a future that Gold conjured up as he delivered his instant classic talk, which went by the title "How Smart Does Your Bed Have to Be Before You Are Afraid to Go to Sleep at Night?" It was a thirty-minute presentation composed entirely of questions, one after another, that Gold asked his Silicon Valley audiences in unwavering succession. The interrogation began with Gold confessing honest bafflement about the motives driving so-called smart environments like those being tested at PARC and MIT: "Why would anyone want to live in an intelligent house? What would be the forces that would compel a designer, or an architect, to create such a thing?"[31]

Reading Philip K. Dick's 1969 novel *Ubik* had raised Gold's suspicions about intricately computerized spaces, even as he participated in PARC's first efforts to build them. The novel's protagonist, Joe Chip, lives in a futuristic apartment that gives a chilling portrayal of smart rooms gone awry. Joe's appliances are each managed and controlled by an interface agent. Dick's fictional interface agents were developed to relieve Joe of the burdens of making coffee, of opening or closing his front door, and so on. The electronic butlers in Dick's tale are, however, owned and operated by a ruthless corporation. Joe must insert coins into the coffeepot and the front door every time he wishes to use them. As long as Joe pays up and complies with the extensive contract he signed for each product, the interface agents politely do his bidding. If he runs out of coins or breaches the terms and conditions, the appliance's friendly-butler persona morphs into that of a corporate lawyer, reciting clauses from the operating agreement and threatening to sue Joe.

Gold figured that this payment dynamic would play out differently in reality. During one of many prescient moments in his presentation, he foresaw that advertising revenues would finance most of this technology. Users would enjoy cheap or no-cost digital services in exchange for opening their screens and their dwellings to perpetual commercial interruptions. He joked, rather frankly, that smart-home owners might receive discounts on their mortgage if they agreed to hang ads on their walls instead of paintings or family photos. After asking the crowd how much of a discount they would need to take such a deal, Gold added, "How about if the advertisements were controlled by your smart house and changed depending on what you were doing during the day?"[32] In-home advertising was just one riff on a deeper theme about control that Gold probed from his first slide to his last. Who actually controlled a smart object, a smart room, a smart home? And what determined its level of "smartness"?

To Pentland's way of thinking, users often gained more agency the more personal information a system gathered from them. The algorithmic calculations informing an interface agent's operations were supposed to reflect the user's desires and preferences. Intelligence was thus a function of consensual, well-meaning surveillance. The Media Lab's smart rooms served at the pleasure of the humans who occupied them, and the rooms' smartness rose or fell with the size of the datasets under their purview. Researchers in Pentland's camp maintained that outfitting homes and offices with sensors made not only for more efficient human-computer interactions; the data collected on individuals within smart rooms, Pentland noted, held a wealth of additional knowledge that merited careful study—a special brand of analysis he eventually labeled "reality mining." In fact, as this research progressed, Pentland's team would become more interested in how smart environments and eventually smartphones could facilitate big-data analyses. Their initial concern with improving user experience gave way to a mounting enthusiasm around the insights, observations, and predictive capabilities that reality mining brought within reach.

But even before Pentland went in this direction, Gold sensed the impending power imbalances lurking in the mid-'90s smart-room experiments; hence his presentation's title question about the fear that might arise upon lying down to sleep in an exceptionally smart bed. He conjured up a bed purported to know you, as it monitored you by means that were not particularly clear, while you were given little means to discern the scope of its knowledge or how it would translate that knowledge into actions directed at you. The bed presented an inversion of the asymmetry problem that Lucy Suchman had detailed in her ethnography of Xerox's photocopiers back in the '80s. Now, rather than the photocopier's inability to sense the human's context, it was the bed that knew more about the situation than the human—the human was effectively left in the dark. Even if the bed catered to the human's desires and anticipated her needs, her comfort would come at the expense of reciprocity. Whenever the bed (like any other smart object) misinterpreted her movements or her speech, she lacked readily available means to correct it. The sleeper would have to adapt to the bed's mishaps when it failed to adapt to her.

Constant tracking and diminished agency aside, Gold intimated that building spaces capable of bending to an individual's every whim might cause more problems than they solved. The Media Lab's smart rooms demanded a staggering amount of electricity and computing power just to recognize a face, a handshake, or a winking eye. To understand even a single gesture, the system needed to process a multitude of other activities. That didn't seem so smart to Gold. In a similar vein, MIT's foremost AI expert, Marvin Minsky, was hoping to create a sport coat lined with sensors and actuators that would "adjust its thermal properties to different climates" in order to maintain one's ideal body temperature.[33] For all the garment's ingenious gizmos, you presumably wouldn't be able to take it to your local dry cleaner. Gold ventured a memorable comparison in his talk to emphasize how myopic such approaches could be: "Which is smarter: awnings over the windows to keep out the sun or a massive interactive, cybernetic cooling system that attempts to keep the temperature of the house within one degree of optimal?"[34] Some of what these experimental devices toiled to achieve had already been achieved with software-free objects. Computerizing those objects with interface agents promised extra functionality and greater precision, but zeroing in on these gains meant reengineering each object in question to suit the technical needs of digital components you wished to add to it. If healthy data flows were what elevated smart things above their "stupid" counterparts, then the mandate to maximize an object's data chops would generally trump other design considerations.

Of the many hats Gold wore, he was a designer at heart. His satirical roasting of smart-home projects appeared to stem from a recognition that this "data first" mentality threatened to eclipse time-honored design principles, which often prized elegance and simplicity over technical sophistication. Prioritizing the latter virtue was fine when it came to desktop monitors and hard drives, but exporting that logic to chairs, toilets, shoes, and mattresses was fated to blunder. Leading technologists working from a narrow definition of intelligence were already creating smart devices that seemed astonishingly dumb to laypeople. Minsky's smart coat, for instance, was more interesting to behold than to actually wear. Gold's questions steadily amounted to an answer that must've crept into the mind of anyone who listened: people would not want every object they owned—nor everything they wore nor every place they frequented—to have a computational look and feel about it. If technology requirements began to dictate the design of things, then whatever good that might come from infusing digital media into physical environments would be outweighed by unwelcome disruptions. "How is an intelligent house different from an intelligently designed house?" Gold had asked. "Given a choice[,] which would you rather live in?"[35] The honest answer was obvious by the time his presentation ended. Connectivity and data might play a supporting role in an intelligently designed environment, but any so-called smart room that prioritized digital operations over physical design considerations was probably a headache in the making, and possibly a nightmare.

Having wrestled with Gold's questions, Weiser and Brown decided to write their paper on calm technology without making any mention of the Media Lab's smart rooms and only one passing mention of the Things That Think (TTT) initiative. This reluctance to affiliate their new design philosophy with these innovations was a telling gesture. Pentland's experiments constituted a logical extension of PARC's Responsive Environments project, and Gershenfeld's shoe computers sprang from the same desires that had motivated PARC's active-badge network. Reflecting on these MIT developments as an outsider seemed to render Weiser more sympathetic to the criticisms his colleagues had leveled at ubicomp devices. When he did acknowledge the Media Lab's work, he generally aired concerns about data privacy that echoed the ones Suchman and her team had advanced in their criticisms of active badges. In interviews, Weiser drew attention to how some MIT's prototypes ran on flows of data that users had little knowledge about or control over. Of the Media Lab's envisioned Body Net, Weiser said: "People may not want to exchange information with everyone they touch. . . . What if their computer sucks information out of me that I don't want to be transferred?"[36] While Gershenfeld professed faith in technical safeguards, as Weiser had earlier, his collaborator Hawley admitted—after describing a TTT concept for a camera that allowed its manufacturers to listen in its customers (as a means to improve the product)—that "probably such complicated things will happen to the messages that are going through the network that we'll cease to understand what's really going on there."[37]

Weiser's reservations in this regard would swell into a broader argument later. For now, he and Brown focused solely on the question of how computerized spaces and smart objects might best present information to people. Even here, the pair saw traces of Negroponte's butler metaphor at play in his protégés' devices. Things That Think projects often asked their users to receive notifications and listen to verbal prompts issued by an intelligent assistant that monitored and interpreted all the data circulating invisibly from their shoes, belt buckle, refrigerator, and so on. While this was certainly a way to free users from attending to screens, Weiser and Brown still preferred the idea (manifested in Jeremijenko's *Dangling String* and in their motorcycle gauges) of presenting data directly to people *without an AI intermediary*, but doing so in a manner that allowed for tacit comprehension.

Hoping to inspire a shift in mindset—specifically among researchers who were already eyeing a post-desktop future—Weiser and Brown circulated their paper "The Coming Age of Calm Technology" far and wide as soon as they completed the draft in October 1996. The gradual expansion of the internet into everyday things, they forecasted in the paper's opening, would "require a new approach to fitting technology to our lives."[38] Innovations programmed to exhibit the immensity of computational power had no place in this emerging frontier, which they predicted would truly arrive by 2020. They referred to the fifteen-year period from 2005 to 2020 as a "crossover point"—a wobbly transition stage during which PCs and early ubicomp products would coexist amid a varied, perhaps incoherent digital landscape.[39] (In this grand scheme, they postulated that web browsers, too, would eventually be outstripped by more nimble and specific gateways to the internet.) Before ubicomp could become the primary mode of digital engagement, technologists would have to start inventing from a completely different premise. Hardware and software for graphical user interfaces had long been crafted to showcase "the excitement of interaction" that awaited the stationary user sitting in front of his monitor.[40] What worked best for video games and web pages was best left to personal computing–use cases.

People would not look at ubicomp artifacts in the same way they looked at the screens of the 1990s. As the internet evolved to play a role in nearly everything everyone did, from driving a car to grocery shopping to going for a run, Weiser and Brown insisted that our digital interfaces "better stay out of the way" in most situations.[41] If lots of objects in many places would each be capable of broadcasting content and capturing data, they should not all be vying to provide an immersive, compelling user experience. "When computers are all around, so that we want to compute while doing something else," the paper continued, "we must radically rethink the goals, context[,] and technology of the computer."[42] Before all else, those who dared to create networks of discreetly connected objects needed to respect the limits of human attention and work within them. Pop-up windows, beeping error messages, and even silver-tongued intelligent assistants were unfit for the dimensions of this paradigm, as were AI-oriented systems like Pentland's that purported to solve for information overload by keeping users on a need-to-know basis and effectively leaving them out of the algorithmic loop. Knowing too little was as enfeebling, and potentially as overwhelming, as having too much to grapple with. Weiser and Brown proclaimed, "Calmness is a new challenge that ubicomp brings to computing" and, ultimately, "for all technological design of the next fifty years."[43]

Calmness designated a sweet spot where computing could be empowering without being overburdening—where the resources of connectivity might yield context and insight without spilling forth to the point of distraction. Media platforms invented in the twentieth century had been trending in the opposite direction. "More often the enemy of calm," wrote Weiser and Brown, "pagers, cellphones, news services, the World Wide Web, email, TV, and radio bombard us frenetically."[44] But this was not true of all technologies. Weiser and Brown returned, as ever, to Heidegger's hammer, Polanyi's white cane, Jeremijenko's *Dangling String*, and the gauges on their motorcycles. What did these things have in common? The two men toiled over the question, together out loud and each in his head, until they extracted a keyword and three basic principles from their scattered examples. The word was *periphery*, and it was, they concluded, a decisive variable that explained why some technologies instilled calm and most did not.

A definition of the periphery accompanied their first principle of calm technology, which stipulated, "A calm technology will move easily from the periphery of our attention, to center, and back."[45] The very thought of an electronic medium that was designed to rest comfortably at the edge of our awareness seemed alien, and it still does. Television shows and the commercials that punctuated them had always aimed to be constantly at the front and center of the viewer's attention. But TVs, like desktop PCs, had an off switch, and they couldn't easily be carried around or operated in places without power outlets. Always-on devices posed a different dynamic, which harbored the threat of distraction but also the promise of being in sync with a changing situation and ready-to-hand when needed. An interface agent couldn't pull this off as well as users themselves could, provided they were equipped with technological aids of another sort: systems crafted to quietly run in the periphery by default. The periphery—that which "we are attuned to without attending to explicitly," as Weiser and Brown put it—remained an untapped resource yet to be taken up in models of human-computer interaction.[46] Things That Think systems were not quite peripheral in this sense; rather, they alternated between being entirely concealed and suddenly appearing to alert their user of something, such as the need to buy milk. Unlike portable PCs, they delivered prompts and notifications to users on a just-in-time basis that corresponded with their location. The system was well attuned to its users, yet the users still had to *attend* to the system because the interface agent still demanded their attention whenever it popped up to instruct them.

Calm technologies needed to be built from a phenomenological understanding of the human body. Each of the senses extended further than the focal points we aim them on. The eyes most clearly saw what was stared at, but also whatever resided around the edges of vision. As the sighted reader directed her gaze upon words, she invariably would hold within view the lines of text above and below, page numbers, her limbs, the desk, and the library floor. She could hear the flip of pages being turned by other readers nearby and feel the ground under her feet, even though neither percept occupied her mind. The importance of these ephemeral sensations to our general awareness went unnoticed. Their significance became clear only when we temporarily cut off the peripheral ties binding our cognitive surplus to the surrounding context. For instance, were the reader in the library to put on her headphones and blast music, the feeling of an unseen friend politely tapping her on the shoulder might elicit a brief shock; had she heard the oncoming footsteps, this peripheral sound would have prepared her faintly for the prospect of such an encounter. Tacit impressions flowing just beneath our consciousness afforded a special kind of relationship with activity happening in the proximity. When we insulated our senses from the periphery, the powers of our concentration stood to lose their contextual footing.

Owing to a fondness for Eastern philosophies, Brown knew that the interdependence of center and periphery had been a tenet of ancient Chinese thought. In the days of the Wei and Jin dynasties (around the year 247), the scholar Wang Bi included "center-periphery" among the four relationships he deemed elemental to human experience. Contrary to the analytical impulse driving European intellectuals like Plato and Descartes, Wang's holistic outlook emphasized connection amid difference, as he and his intellectual circle refused to divide the world into separate entities. Isolating discrete parts from a sense of the whole had always helped spur theoretical reflection and the scientific method, but these habits of the Western mind can turn counterproductive outside the study and the lab.

An application for this Old World idea clicked into place one afternoon when Brown and Weiser took their seats in the PARC auditorium to watch a skit put on by Eric Saund. Saund walked on stage with a pair of toilet-paper tubes taped around his eyes, like a set of protobinoculars. He wished to dramatize his idea for a new product he was calling Docufinder. At his side lay a carefully arranged mess of paper. The Docufinder would use infrared laser and computer vision to track every single sheet of paper piled on a desk, keeping tabs on each one even as it got buried in an unorganized stack. Saund, with his costume, played the role of the Docufinder. A colleague entered stage left and pretended to be an office worker who couldn't find the piece of paper they needed. Saund directed his tubes at the pile, scanned it for a second, then shined a laser pointer on the document his colleague was searching for. Once the waves of appreciative laughter from the crowd had died down, it occurred to Brown that toilet-paper binoculars could serve as a nice metaphor to help him and Weiser articulate the root cause of the eye-glazed exhaustion people felt after surfing the Net. Soon, the image found its way into their coauthored writing: "Wearing cardboard tubes is much like living in the digital age. . . . Today's digital technology and delivery mechanisms tend to flatten and push all information to the center of our awareness . . . effectively cutting out the periphery."[47] The array of online applications running on a typical web user's computer generated a unique brand of information overload largely because each application made overlapping appeals for the user's undivided attention. Clicking between an open email, an excel spreadsheet, and a chat box, not to mention from one website to another, felt like trying to navigate things one toilet-paper-tube glimpse at a time. Overload ensued—not due to the volume of content, but because the peripheral sensations weren't there to offer context. Information abundance was a problem only when information technologies reflected a diminished view of our attentional capacities.

Eager to push their fellow researchers in the opposite direction, Weiser and Brown extrapolated calm technology's second principle. "Technologies encalm as they empower our periphery. . . . A technology may enhance our peripheral reach," they wrote.[48] The periphery, then, wasn't simply a holding area for storing excess information until one needed to attend to it. The periphery was a powerful form of attention in its own right. Like a radar, it continually processed a little bit of information about a lot of things. Ubicomp designers could regard the periphery as a zone through which vital knowledge emerged. In addition to staying out of the way, an advanced calm technology might also "bring more details into the periphery"—but not just any details.[49] The goal should be to cast into our periphery only data that relates to the "something else" we are doing while computing. When a motorcyclist was riding down a curvy hill, a display showing his current speed was bound to be more important to him at that moment than a newsfeed listing the day's top stories. Having a precise miles-per-hour number at a glance provided a useful means against which to check his intuitive sense of velocity. This was a data point that strengthened his attunement to the motorcycle and added context to inform his quick decisions on the road. Speedometers predated ubicomp, of course. The novel thing now was that, with ubicomp's impending growth, any object could stand in for the motorcycle and any information in place of speed. Instead of a numerical gauge, the status of a given metric could be even more subtly communicated by a spectrum of color changes, noise levels, sound cues, or haptic feedback. Moving through the world with a digitally enhanced periphery, supported by all manner of intelligently designed systems, might better inform our ability to interpret situations in all facets of life.

Calm technology would neither replace nor entirely displace the web, Weiser and Brown made sure to point out. The web was absolutely an incredible set of toilet-paper tubes. One after another, electronic documents from the world over could spring into the user's focused visual field. The web's creators had utilized the internet's connective prowess to assemble loads of disparate content within reach from any single screen. Still, digital networks remained capable of far more than the part that web browsing drew upon. Snippets of the online universe could be selectively linked to certain objects and not others. Computational media could be paired with physical materials on a local, situational basis—as Jeremijenko had paired data about PARC's Ethernet traffic with the motor and red cord hanging near Weiser's office. Whereas the web's shape-shifting magnitude tended to leave novice users dizzy and discombobulated, a technologically-enhanced periphery would foster a deeper sense of "locatedness," Weiser and Brown concluded. This heightened connection to the nearby world furnished their third and final basic principle: "The result of calm technology is to put us at home."[50] And with that, they circled back to Xerox PARC and described a few projects around the building that exemplified their calm philosophy.

The paper ended where Weiser had begun, with a techno-philosophical ode in praise of Jeremijenko's *Live Wire*. Despite the newer prototypes that were garnering popular acclaim, Jeremijenko's artwork still appeared closest to the spirit of their vision. The age of calm technology was coming, the paper's title assured readers. But its authors weren't exactly sure where to go next. While Weiser remained ambivalent about the Media Lab's recent uptake of ubicomp, he was keen to find collaborators at MIT.

9

Smarter Ways to Make Things Smart

PHIL AGRE HAD JOINED MIT's doctoral program in computer science eager to work with the nation's top AI researchers. His enthusiasm gave way to questions; his questions led him to have serious doubts. When recounting his time there, Agre wrote, "So here I was in the middle of the AI world—not just hanging out there but totally dependent on the people if I expected to have a job once I graduated—and yet, day by day, AI started to seem insane."[1] Agre noticed patterns playing out in conversations around the various MIT labs tinkering with AI and at conferences in the field. The peer pressure to hack together "what works" effectively stifled any impulse to stop and talk about what made the most sense to build. The discipline's inclination to oversimplify amorphous concepts in the face of immense technical demands (captured by the Media Lab slogan "Demo or die") seemed to drive much of what Agre observed: "They would, by and large, rather get it precise and wrong than vague and right."[2]

Throughout the 1990s, Agre found examples of willful folly in the AI community's seminal texts. He rebuked Allen Newell and Herbert Simon's 1963 paper "GPS: A Program That Simulates Human Thought," as he emphasized how shortsighted the authors were when they decided to model the world as a vast multiple-choice question that each person answers one step at a time. "The environment is reduced to the discrete series of choices that it presents in the course of solving a given problem," Agre lamented.[3] The world was not always—certainly not only—a problem to be rationally figured out. That such notions to the contrary were entertained and even lauded within the AI tradition indicated to Agre just how deeply the field was rooted in Cartesian ideals. This reductive sense of the environment stemmed from a diminished appraisal of the body. Descartes's unforeseeable influence on AI research, Agre contended, lay primarily in the legacy of the philosopher's relentless effort to "partition functions between body and mind."[4] The body in Descartes's schema was a machine. And though it was a machine "made by the hands of God," the body did not harbor any sort of knowledge. It was merely an instrument of the mind or soul, which was the real locus of intelligence.

A similarly disembodied conception of intelligence propped up Alan Turing's famous "imitation game" proposal, in which he set the criteria by which AI specialists would assess their prototypes for decades. Put a man in a room with a computer terminal, Turing ventured, and have him converse in typed messages with two hidden interlocutors: one human and the other a computer program. If the man cannot correctly guess their identity over half of the time, then the computer program must be deemed to have successfully approximated our humanity.

In November 1996, a journalist for *Computerworld* magazine alluded to the Turing test in an email he sent to Weiser. The journalist, Mitch Wagner, had just interviewed Nicholas Negroponte and Marvin Minsky at MIT, as well as Microsoft's CTO Nathan Myhrvold, in order to gather their thoughts for an upcoming article about the future of AI. Wagner figured some remarks from Xerox PARC's chief technologist would round out the piece. He offered to visit Palo Alto to hear Weiser's take on the question of, as Wagner put it in his message, "how likely it is that we might create a machine that has a human-level intelligence—or appears to."[5] The three men with whom he had already met each supplied positive if measured sound bites that corroborated AI's significance as a technology that would, in perhaps ten to fifty years, change everything, and probably for the better. In reply to Wagner's interview request, Weiser wrote back, "No thanks. . . . AI is not very interesting—I'm against it."[6] The journalist found a fourth proponent elsewhere.

Whenever Weiser found AI interesting enough to mention, he opposed it on grounds similar to those covered in Agre's more detailed criticisms. The men had each been influenced, independent of each other, by similar sources. (Agre drew heavily on Martin Heidegger's philosophy, too, and he revered the writings of his friend Lucy Suchman.) Both Agre and Weiser felt alarmed by the smart systems being developed at MIT and especially by its scientists' candid reverence for disembodied life. The Media Lab's senior luminaries saw no essential impasse at play between microchips and flesh. They maintained that the body's contributions to cognition could be replicated with the right gizmos. Minsky told *Computerworld*, "If you could understand how to represent the important elements of the human personality, then eventually you could build a scanning machine and extract the essence of the person—the intellect, the theories, the ideas, the way of thinking—and put it in another piece of hardware that would last longer than human bodies."[7] Negroponte implied that future AI devices, equipped with facial recognition and emotion detection, might simulate a gifted elementary school teacher's ability to understand a child.

Weiser began flying to Massachusetts more often in 1997, in between his travels to the industry's mega-events such as the World Economic Forum, where he presented alongside would-be Google CEO Eric Schmidt, and the Living Web conference, where he listened to Jeff Bezos and Bill Gates pontificate about emerging models for doing business online. Conversations at these events revolved around web browsers and the opportunities they held for e-commerce, information search and retrieval, entertainment media, and instant messaging. Weiser remained convinced that popularity of websites would eventually plateau as personal computing waned. The internet would take on new forms once mobile, wearable, and embedded devices became widespread. Exactly which forms it should take was still an open question, as were questions concerning AI and the role it ought to play in a ubicomp future. The winning answers would most likely emerge from MIT, or at least pass through it.

When Xerox executives decided to join the Media Lab's sponsor program, Weiser jumped at the chance to be PARC's official liaison. He wanted to forge better relationships with the younger researchers there who were toying with various ways of wirelessly connecting everyday things. He needed to convince at least some of them that AI had no place in their experiments, even though he knew AI was all the rage there.

Weiser may have felt like a walking antithesis to some of the projects he saw on display at the Media Lab's events, but he was antsy to identify a basis for collaboration. The best version of ubicomp was a staunchly modest one—Weiser realized even more clearly after his and Brown's paper on calm technology. Weiser's ideal computerized space might be packed with smart objects, but none would do much more than convey information to people in their periphery. The purpose of adding computing to things, he now believed, was to make them more usefully expressive. In the style of Natalie Jeremijenko's *Live Wire*, Weiser hacked together a small fountain outside his office whose water levels rose and fell every day in accordance with Xerox's stock price. Whimsical and satirical, the fountain nonetheless provided a second illustration of calm technology's operating principles: a commonplace physical substance (water, in this case) could tacitly communicate the meaning of digital data in real time. The built environment could be programmed to present all sorts of computational insights to people in ways that didn't require them to stare at any screens. Between the Things That Think prototypes and Alex Pentland's AI-powered smart rooms, the former seemed more amenable to Weiser's agenda.

Inventions underway in the Media Lab's Things That Think group, led by Neil Gershenfeld and Michael Hawley, installed just enough computational awareness into things to render them capable of performing a couple tasks without human assistance. For instance, with the insertion of an electronic tag, even a chicken breast could be made to "think" on its own. Once placed in the oven, this tag in the meat could "automatically set the correct temperature . . . and turn off the heat once cooked to taste."[8] Using tags and sensors to ferry data from thing to thing promised to relieve people of the need to press buttons with their fingers, and the information these items captured could be transmitted instantly to a third party. Gershenfeld would write, "In a TTT world, the medicine cabinet could monitor the medicine consumption, the toilet could perform routine chemical analyses."[9] While these applications were geared toward admirable ends, Weiser worried that the means by which they achieved those ends would effectively prioritize automation and data mining over transparent communication. Gershenfeld and Hawley's prototypes had features in common with those Weiser had presided over earlier in the decade. PARC's Responsive Environments project enabled the building's HVAC system to adjust room temperatures to suit each staff member's preference—the active badge they wore was like the TTT tag in the chicken breast.

Weiser's sense of "how much is too much" had changed since then. None of the TTT applications were so disconcerting in themselves, of course. But the prospect of many more smart objects forged in this fashion—all designed to function without human involvement—conjured a very different future than the one Weiser had envisioned. A TTT world might relieve people of the burden to pay close attention to the things around them; smart objects would act on their own rather than enhancing a user's situational awareness and leaving it for her to act. The latter was Weiser's highest priority. Marginal gains in efficiency or accuracy won through TTT-style automation were not worth all their little encroachments on an individual's agency. An object could be programmed to wiggle, hum, light up, and alter itself in other ways, Weiser countered, so long as its dynamism served to tacitly convey relevant information to the user. He argued that the payoff of any so-called smart object ought to be "our increased ability for informed action."[10] The Things That Think group was inching toward automating all kinds of activities; so long as the things they built did the job well, the research group did not seem too concerned about whether the human beneficiaries were informed or not. But Weiser had reason to believe that all these little automations might set some unintended consequences into motion.

As Gershenfeld and Hawley's initiative gained momentum, an even more ambitious smart object was taking shape at a Dutch shipyard, and it offered another version of a "TTT world." The cofounder of Netscape had a new project in the works he called Seascape. Jim Clark's rush to take his web browser company public was, it turned out, in no small part motivated by his yearning to finance the sailboat of his dreams. Shortly after Netscape's historic IPO, Clark hired Europe's best yacht maker and flew a team of his best software engineers over to Holland. Not only would it be the world's tallest sailboat, but Clark wanted every aspect of the vessel, from the sails to the light switches, to be governed by computer programs. He even wished to monitor the yacht's status from his office in Silicon Valley and, if he desired, steer it for a lap around San Francisco Bay using a remote control on his desk.[11] In order for this to work, all the yacht's mechanical systems needed to be outfitted with tiny sensors that "measured everything Clark could think to measure."[12] Streams of data about the engine pressure or the DVD player in the cabin, for instance, were all transmitted to programs running on the yacht's souped-up computers.

The few crew members with whom Clark had shared his password enjoyed access to "God Mode": an omniscient glimpse of the ship's every detail and the power to alter anything with the right keyboard command. Clark's software engineers had joined the project in hopes that *Hyperion* (the boat name Clark settled on) represented a nautical harbinger of things to come on dry land. They suspected that *Hyperion* might prove to be "the first Home of the Future," according to Michael Lewis when recounting his time on board observing the crew.[13] Further, "the computer would permit the owner to enter into a new, fantastic relationship with his dwelling. . . . One way of viewing *Hyperion* was as a test of the technology."[14] And the test didn't go so well. Lewis chronicled an array of scenes dramatizing how the yacht's digital operating system rendered it beyond the control of its captain, its crew, the engineers, and even Clark himself. Unlike *HAL*, this ship's computer had no evil intentions; rather, the complexity of its code left it riddled with enigmatic bugs that yielded dangerous surprises. As the ship's chef prepared a meal in the kitchen, her table suddenly rose toward the ceiling, then crashed to the floor—"dishes stacked on the table flew in eight directions."[15] The ship's engine developed a habit of turning itself off in the middle of the night, leaving it to idle in the middle of the Atlantic Ocean, and no one on board ever figured out why.[16]

Malfunctions of this sort were the least of Weiser's misgivings. Programmers would improve the applications over time; any features that remained too finicky or hazardous could be removed. Even if all the AI-infused objects aboard a rich man's yacht or inside the average home behaved like they were supposed to, a vast majority of people would still be subjected to some version of God Mode. Appliances, furniture, and mechanical systems would carry out your best interest, whatever the software calculated that to be. But unless you had the access and ability to program them, environments that thought and acted on your behalf would prescribe only limited opportunities for you to intervene. Weiser maintained that sensors and software, no matter how much convenience they stood ready to deliver, should never entirely relieve humans of the responsibility to judge a situation. "People don't like the idea of machines running their life, carrying on a dialogue between themselves about their owner," Weiser told a reporter.[17] He echoed this sentiment in meetings with his colleagues, when he would say, "I don't want to argue with my car about where I want to go."[18] It wasn't that the machine would have a mind of its own beyond all human control; it would just be designed to make decisions for users in a manner its programmers, not its driver, deemed best.

In the latest rendition of ubicomp that Weiser was pitching in his travels to MIT, he contended that smart objects ought to communicate in plain view. His sole purpose for weaving computing into things was to make relevant insights ready-to-hand for people in a variety of settings. The information made actionable by ubicomp systems could enrich human decision-making; no algorithms or interface agents were tasked with making decisions in Weiser's designs. Voicing his reservations about excessive automation, Weiser insisted that computerized spaces should only help *us* think. To do so, the data they processed needed to become "part of the informing environment, like weather, like street sounds," rather than sequestered behind God Mode.[19] "We become smarter as we put our roots deeper into what is around us," Weiser intimated during his Media Lab visits, surveying the faces for a pair of receptive eyes.[20] Instead of trying to automate the user's every need, Weiser believed, technologists should computerize things simply as a means to heighten awareness among people who were present at the scene—neither the system itself, nor coders who oversaw it from afar, should exercise any direct control over those in the vicinity.

Hiroshi Ishii, a young Japanese computer scientist, was the most kindred spirit Weiser found at the Media Lab. Though Ishii's research was supported by MIT's Things That Think consortium, the technologies he worked on rarely won mention in news stories profiling the Media Lab. His inventions weren't the sort of devices people had grown up fawning over in sci-fi movies. Ishii would eventually receive the highest honors in his field, including the SIGCHI Lifetime Research Award, but his early work remained obscure relative to that of his hotshot colleagues down the hall. The project that might make him a star had thus far barely garnered academic notice. Weiser learned of him in 1997 when a rough draft by Ishii and his graduate student turned up in Weiser's mailbox. Organizers of the ACM CHI Conference on Human Factors in Computing Systems had sought Weiser's opinion about whether to accept a paper Ishii submitted titled "Tangible Bits." That paper (now famous in technical circles) almost failed to reach any audience whatsoever. Weiser's advocacy for "Tangible Bits," Ishii claimed, "rescued [it] from the brink of rejection."[21] Right after Weiser finished reading it, he dashed off an email to Ishii. "That email became my greatest treasure in life," he later said, in a testament to how crucial Weiser's confidence in him was.[22] Weiser had confided to a then-struggling Ishii: "This is the kind of work that will characterize the technological landscape in the twenty-first century."[23]

In "Tangible Bits," Weiser saw something like a sequel to his own *Scientific American* essay from 1991, in which he had defined ubicomp as an ascendant paradigm that would gradually overtake desktops and laptops. Extending Weiser's critique of the personal computing model, Ishii's paper dared to question the computer-human interaction community's most heralded innovation: the graphical user interface, or GUI. Deference to this acronym and the ideas for which it stood—its reliance on virtual windows and a mouse-controlled screen—had been halting the progress of ubicomp. Weiser's vision lacked a GUI-like notion, a pragmatic framework clear enough to unify comparable R&D initiatives. Ishii devised a new acronym to pitch the CHI crowd: TUI. His "tangible user interface" sought to make bits of online data and their meaning more palpable to at least four of the five human senses.[24] (Taste was set aside.) All physical matter at rest in Ishii's lab was a communication platform awaiting its electronic activation—"not only solid matter, but also liquids and gases."[25] Beyond typing and clicking, Ishii hoped to give people ways to engage with the internet by taking hold of "everyday graspable objects (e.g., cards, books, models)."[26] To access or generate online content, the user of Ishii's TUI cards might simply point one at a building or the book in her lap, rather than sitting down to a PC and opening up a web browser. Ishii, like Weiser, looked forward to a world in which everyday things were "seamless[ly] couple[ed] . . . with the digital information that pertains to them."[27] Jeremijenko's *Live Wire* at PARC was a superlative TUI in this sense, and Ishii championed her project alongside several in his own lab. His ambientROOM, for instance, deployed "ambient light, shadow, sounds, airflow, and water flow as a means for communicating information at the periphery of human perception."[28] Whereas desktop interfaces revolved around digital representations of office items, in Ishii's experimental room, an actual desk filled with sensors played the roles of screen, keyboard, and mouse all at once.

To illustrate, Ishii gave the example of a toy car that his lab had made into a TUI. Whenever he placed the car on the desk, that simple action projected onto the desk a detailed infographic presenting statistics about recent traffic on the car's web page. For a more ambient display, Ishii could place the toy car beside a speaker in the ambientROOM and the speaker would broadcast the sound of raindrops—a drizzle when the web-page traffic was light, a downpour whenever it picked up. Finally, if Ishii preferred silence but still wished to be in the know, he could place the toy car near a little water tank he had rigged up to ripple with the flow of the web page's activity. His aquatic contraption made the water's movement faintly visible throughout the room: "Light projected onto the water tank from above casts a subtle but poetic images of ripples on the ceiling."[29] The significance lay not so much in the toy car itself as in the process the car exemplified. Almost anything could transmit live digital information to almost anything else, and its message could be variously delivered to the ear, the eye, or the reading brain. This shift "from the GUI of desktop PCs to TUIs," Ishii's paper forecasted, "will change the world itself into an interface."[30]

Together, Weiser and Ishii hit the international speakers' circuit, sharing the stage at industry conferences across the US. It was at a relatively small festival in Europe, Ars Electronica, where Weiser spoke out in praise of bodies. The more he listened to Ishii talk about making computing tangible, the more he realized that the body was a crucial missing piece of the calm technology philosophy he had outlined with Brown. Ars Electronica brought some one thousand artists, scientists, and designers to Linz, Austria, a winding medieval city of white-, pink-, and yellow-stone walk-ups, halved by the Danube River and hugged on all sides by rolling hills. The festivalgoers congregated inside a massive glass hangar called Design Center Linz for five days of "grand visions," "unusual prototypes," and "rousing concerts."[31] Weiser could've delivered on all three during his allotted time, but he stuck to conveying a vision of the body he hadn't discussed publicly before.

"Not every augmentation of our bodies will work," Weiser said before the crowd.[32] It wasn't the kind of claim you expected to hear from the so-called father of ubiquitous computing. But Weiser wasn't exactly the same guy who had, over the previous years, posed for magazine cover photos always with some new device in his hand. He came to Linz without any hardware to flaunt, and his prepared remarks made almost no mention of computers. What he brought instead was an itch to share insights from his latest philosophical foray. In the writings of the French philosopher Maurice Merleau-Ponty, Weiser had found another anti-Cartesian theory of body and mind that built upon the core ideas he'd internalized from Heidegger and Polanyi. Merleau-Ponty's books, filled with expressions like "the body understands," pushed theses that ventured far afield from AI's body-agnostic sect. Merleau-Ponty gave limbs and skin a high status in the life of the mind; the body he invoked was "the body no longer conceived as an object of the world." Rather, the body was for him "our means of communication with [the world] . . . the world no longer conceived as a collection of determinate objects, but as the horizon latent in all our experience and itself ever-present and anterior to every determining thought."[33] The human body—whatever form it took—was a medium, not a machine. The world, as one's mind configured it, necessarily drew from the flow of stimuli regulated by the body's perceptual capacities. Our senses did not process information; collectively, they proffered intuition ("the horizon latent in all our experience"). Attempting to augment the body with sensors and wearable screens that obstructed one's senses was, from a phenomenological viewpoint, like building a pair of mechanical wings to help fish fly. It might achieve the engineer's objective, but only at the cost of losing touch with the elements that furnished our being-in-the-world.

Barely rewording the Frenchman's adage "The flesh is at the heart of the world," Weiser told his Linz audience: "At the core of everything proximal is the body."[34] The word *proximal* carried a dual meaning here: it encompassed the things situated in our vicinity as well as our tacit impressions of those things. The sensory-engaged body, Weiser continued, never experienced separation from the world—the body felt attuned to things that bypassed consciousness and nevertheless furnished the cognitive landscapes where conscious reasoning played out. This state of effortless connection was, in Merleau-Ponty's account, more natural to humanity than the alienated postures that twentieth-century critics of modernity had scorned. By tying our being so tightly to our body, Merleau-Ponty asserted that entanglement was not some kind of enlightenment we needed to think our way toward. Entanglement was a casual matter of fact: "The world is wholly inside and I am wholly outside of myself," he wrote.[35] It was only when we fixated habitually on the explicit contents streaming through our minds that we could fall privy to delusions of isolation. If computers were tools to amplify heady notions of cognition—a bicycle for our minds, as Steve Jobs put it—then what did that mean for the body?

If you bought into Merleau-Ponty's premises about the body as Weiser did, then a whole lot of new media began to seem utterly misguided. "Current digital technology suffers from a painful lack of connection to the deepest foundations of being human," Weiser told the crowd.[36] Two-dimensional programs on screen, distant yet irrevocably explicit, compelled one's attentive mind while minimizing bodily involvement. Of what enduring good was any technology that short-circuited the robust, live transmission between one's body and the world?

Totally automated environments such as Clark's smart boat harbored grave phenomenological consequences that were clearly bad for bodies. To the extent that software administered from elsewhere materially dictated what could and could not happen in a place, that locale got rendered more and more into "a collection of determinate objects" that no longer respond to physical intervention. All bodies without the right computer or the right password were confined to the role that a programmer scripted, and they remained unable to mitigate problems stemming from any bugs lingering in the program.

Proponents of mobile computing, on the other hand, fancied they were honoring our embodied nature by developing systems that could be used on the go—especially when these featured displays could be controlled via touch and software that could listen and speak back. But the body's fundamental importance, Weiser gathered from Merleau-Ponty, amounted to much more than its capacity for locomotion or the various input modalities it brought to computer-human interactions. Any apparatus that even slightly impaired the senses was all the harder to justify when the user was moving through physical spaces, where her body's ability to "take in and act out the periphery" was paramount, Weiser insisted.[37]

As a medium for noticing nuance and understanding changes in the proximity, the sensory-engaged body could not be replicated by any concoction of microchips and code. Weiser suggested that the body's grasp on the proximal ought never be obstructed. To augment the body's peripheral reach, to heighten its sensitivity to the proximal, technologists should before all else leave the body unencumbered. They should instead do what Ishii and his team were doing, which Weiser had described as "awakening computation mediation into the environment."[38] The best way to augment the body *was to augment the material world*—to add meaningful, dynamic, sensory nuances to objects that people could become attuned to as they went about their day.

Scaling technologies like Ishii's TUIs to offices, schools, homes, and many other settings would, Weiser said, inaugurate "the age of the periphery" wherein our bodies would "extend everywhere."[39] Not a geographical extension in the vein promised by telepresence, but rather the extension of our perceptual capacity to receive a wider swath of stimuli in a manner that we are most adroit at making sense of it. Arrays of information would be transcoded into sensory cues tailored to jibe with the body's intuitive dimensions. In contrast to the constant attention involved in surfing the web, the future Weiser spoke of would entail "living with, powering ourselves by, the constant periphery."[40]

With that, he concluded and turned the stage over to Ishii, and Ishii showed the audience images of everything he had in the works at MIT. The session ended in applause more rousing than any they received following their presentations earlier that year at corporate American gatherings. Those crowds—every bit counterpart to the Ars Electronica conference-goers—remained much more invested in the Media Lab's other ventures.

It was a shame they never insisted on a more adventurous moniker. *Perceptual computing*, *wearable computing*, *affective computer*—these and other technical descriptors used by Alexander Pentland, Rosalind Picard, and a troupe of graduate students to label their team efforts did not begin to capture the peculiar, techno-chic *art de vivre* exhibited in the photos they took to showcase their technologies. Each showed the researchers together on MIT's campus striking a stately pose. Weighty fanny packs sagged around their jeans at the waist. Wires traveled under and over their jackets and hoodies, leading to the hardware on their heads. Covering their eyes were dark wraparound visors of all different cuts and sizes, except for one, whose otherwise conventional eyeglasses had a three-inch piece of plastic stuck to the left lens. The group's carefully unsmiling mouths, faint goatees, and postured coolness, bore some resemblance to the era's famed pop bands. Within certain technical circles, their research could garner that level of attention.

By fall of 1997, Pentland and company had largely moved on from the smart rooms he wrote about the year before in *Scientific American*. They retained all the room's sensors, cameras, and facial-recognition software. Instead of bringing users into their lab, they strapped these technologies to the users' bodies. When a user wore the room's devices on her person, any room she entered could be casually graced with AI functions and, in turn, enrich the team's algorithms with extra heaps of data. And not just more of it, but also new kinds. The smart room in Pentland's lab gleaned only the outward manifestations of personality. These wearable iterations penetrated below the surface of body language, speech, and appearance. Sensors on the body could monitor the subject's vital signs as she went about her day, and even while she slept. The group's interest with bodily data was not entirely medical in nature. The body's metrics, as they seemed to regard them, were the nearest thing to a window into the user's soul. The more technologists knew about the user, the better odds they had at programming interface agents to serve in her best interests. In any event, regardless of the software's performance, this new arrangement generated a bounty of information, so personal and specific to each person, that had lain untouched by other forms of computing.

The world got to see everything the Wearable Computing group had

MIT Media Lab researchers showcasing their head-mounted display systems. Photograph © Sam Ogden. Used by permission.

in the works over three days that October. Monday's and Tuesday's events catered to the academic sector, with a packed lineup of concurrent panel sessions on all technical elements. Wednesday was the day of the Wearables gala. The 1,400-person guest list included camera crews from ABC and the BBC, and reporters from the *New York Times*, *Boston Globe*, *Business Week*, *Newsweek*, and *Fortune*, along with the standard assortment of notable executives and company liaisons who flocked to the Media Lab's sponsors-only affairs. Preparations for this round of festivities went even further over the top than usual.

Weiser had been on hand for the presentations during the first two days, and on Wednesday morning he took a seat in MIT's Kresge

Auditorium—the same venue where he and Negroponte quarreled on stage five years earlier. Outside, the Cambridge sky was dull with clouds like it had been then. Weiser could only feign excitement as the gathering audience eagerly read over their glossy programs. A butler would not be hosting the proceedings this time. No, for the occasion, the Media Lab had secured the actor who played *Star Trek*'s Mr. Spock (Leonard Nimoy) to introduce each speaker. First, there would be Negroponte's opening remarks, then the headlining talks by Pentland and Picard, and some brief demos by a few others. Then, there would be a fashion show.

Already weaving through the crowd were MIT researchers decked out in electronic outfits of all sorts. Jennifer Healey, a PhD student of Picard's, described her look—silver face paint accented by a metallic headband wired to a CD player—for an inquisitive reporter: "If I'm calm," said Healey, pointing to her headband, "it will play 'Scarborough Fair.' If my skin is in a high state of arousal, it will play 10,000 Maniacs."[41] This was one of several "biosensor" applications she and Picard were developing to give the body's stress responses a place in computing's feedback loops. Among the other graduate researchers showing off their wares were Steve Mann and Thad Starner. Glitz and glamor aside, the gala's costume theme demanded nothing atypical from them. Both had been dressing in gadgets on a daily basis for years. Mann's commitment started back in high school, when he got the idea to stuff his backpack with bulbs, cameras, batteries, and a 6502 computer—of the kind used in the day's Atari and Nintendo gaming systems—while doing his best to repurpose his head into a small radio tower.[42] His current setup had evolved considerably, as the gear clipped mostly around his belt now and his head-mounted display resembled oversize sunglasses. Starner had begun wearing full time in 1993. Electronic eyewear was his passion as well, though he favored a lower-profile design than Mann. Starner attached a thin strip of hardware over the lens of store-bought glasses and called his pair "The Lizzy."[43]

With Pentland and Picard's support, Starner and Mann had initiated the Media Lab's smart-clothing project in 1992, but it was only lately that their advances in software made their hardware an enticing proposition. These latest breakthroughs elicited favorable reactions during the scholarly proceedings of the past two days, a reception that must've made them appear all the more fraught in Weiser's eyes. Each of these body-hugging gadgets contradicted the phenomenological view of the body he had championed alongside Ishii at Ars Electronica. The Wearable Computing group sought to collect information from the body while filling the senses with multimedia. A growing roster of applications they once ran on desktops was running now on their wearable interfaces.

At the core of this recent progress lay a program called the Remembrance Agent. In its first iteration, the Remembrance Agent dealt exclusively with text: it read over everything the user typed onto the screen and repurposed those typed words into search terms for locating any files in the computer that might prove relevant to the user's present activity. Whenever Starner typed notes about an article or lecture, the program would almost instantly, in the corner of his screen, present him with possibly related documents already stored on his database. Nominally useful at best, this was just a starting point. The Remembrance Agent became more powerful when fused with other technologies around the Media Lab, such as Pentland's systems for facial recognition and word spotting. A pair of smart glasses equipped with all three capabilities could identify various sights and sounds in a room, then use this audiovisual input to comb for valuable information on the user's behalf. Like an executive assistant, the glasses could identify oncoming faces and whisper their names into their wearer's ear or display them at the edge of her vision. Using the word-spotting feature, the glasses could listen for the utterance of certain words and automatically retrieve corresponding content. "Whenever someone mentioned 'the Megadeal contract,' the software could project Megadeal's finances onto the display in your glasses," Pentland wrote, conjuring a hypothetical scene of the tech in action.[44] Evidently, the thought of filling one's idle moments with such infographics—life as a long succession of PowerPoint slides—was not so insidious to his readers. Beyond this, the research Pentland and Starner had presented at the previous day's scholarly gathering hinted at how the technology might unburden people from having to learn complicated skills or grow their intuition through repetitive practice. One such application they demoed, a baseball cap with a camera attached to it, could parse American Sign Language with 98 percent accuracy. Another work in progress named "stochasticks" sought to help billiards players plan and take better shots.[45] It, too, featured a computer-vision algorithm to classify objects on and around the pool table. Then it determined which shot the user should take before rendering in the lens of his eyewear the precise angle and point at which the ball should be struck. Their talks basically implied that the mental side of billiards (or, really, any comparable game) could be reduced to a simple matter of following standard pop-up instructions.

The pre-gala mingling of expensive suits and cyborgs-in-residence finally subsided; everybody found their seats, and Leonard Nimoy did his thing. After Negroponte wrapped up his welcome speech, Pentland and Picard gave the audience a nickel tour of their favorite projects, which they fashioned as a sneak peek inside the lab's "wearables closet."[46] The items Picard showcased revealed the surprising depths to which the Wearable Computing group could probe a user's psyche. Be it jewelry, garment, or facial accessory, many of their prototypes toggled fluidly between the audiovisual and the affective. Tiny cameras and microphones linked with sensors to process images and speech, while other sensors traced the wearer's emotional states and affective patterns. The whole ensemble could operate in concert, too. The constant, real-time emotion detection that Picard's research added to this mix powered eyewear and earrings alike to capture the ebb and flow of one's mood, interest level, or stress response. These affective measures—inferred from signals like heart rate, skin conductivity, body heat, and so on—then determined how each device would go about serving the user from moment to moment.

For music lovers, there was the CD-player contraption that selected songs to match the user's current physiological vibe. (This was the

A billiards player using the "stochasticks" system developed by MIT Media Lab researchers. Photograph © Sam Ogden. Used by permission.

device that Picard's face-painted protégé was wearing.) For midnight walkers, Picard touted Safety Net—a line of wearables that could detect fear, whereupon the software would command the camera to snap a wide-angle photo of the user's viewpoint and electronically send the photo, along with the exact location and a designation of the user's "fear state," to a group of confidants with whom he wished to share such moments. Two other applications, Memory Aid and the Intelligent Web Browser, seized on the user's attention and interest. The former was programmed to have the camera record whenever it deduced the user's attention had wandered, while the latter was designed to seize on moments when online content piqued the user's curiosity by stoking his interest further with additional, related content. The AI community, Picard observed, traditionally ignored users' emotional states. She, Pentland, and their graduate students had introduced a brave new conception of AI, rooted in the pairing of wearable sensors and the affective data they collected.

It is unclear whether or not Weiser hung around for the fashion show that followed. Set to a thumping "neo-disco" beat and billed under the theme "Beauty and the Bits," the show brought to the stage professional models attired in more than forty distinct computational outfits, each of which had resulted from collaborations between MIT techies and aspiring designers at top fashion schools in New York, Milan, Paris, and Tokyo.[47] The camera crews in attendance sprang into action. For a man who took notes almost everywhere he went and made a point of saving them, Weiser scribbled none of his usual marginalia asides or exclamation-point-studded ideas. All that remained was his ruffled copy of the day's program.

Between the desert towns of Fernley and Ely, Highway 50 cut through Nevada's high country with the merest specks of civilization dotting its 287 miles. A reporter for *Life* magazine once described it as "the loneliest road in America, with no points of interest"; he advised readers never to drive it, on account of the boredom and a genuine risk of getting stranded amid its barren sands and snow-covered peaks.[48] But some Nevada state officials found a shard of poetry in the reporter's slight. What the highway lacked in modern services, it more than made up for in trance-inducing imagery and sublime stillness. New road signs were installed, and a marketing campaign ensued: Nevada's stretch of Highway 50 became, proudly, "The Loneliest Road in America"—a weekend adventure to cross off one's bucket list. On most weekdays it was desolate.

No one was around to witness Weiser roar across a 120-mile stretch in a single hour, his receding hair blowing about in every direction as he screamed along to the Jimi Hendrix songs blasting through his speakers while his colleagues worked back in California. It was mid-March and the western air blew chilly even at a standstill, to say nothing of its effect on motorists speeding in a convertible with the top down. But Weiser, a proud owner of a Mazda minivan ("good for hauling drums around"), couldn't let his time with a BMW Z3 expire without pushing it to the max.[49] A week with the Z3 was Weiser's prize for pitching the winning idea at a recent engineering retreat that BMW had convened on "the car of the future." Knowing nothing of automotive design, he had as a lark just told the company's chairman to "fire half their mechanical engineers and replace them with software engineers."[50] He still didn't know why the chairman had picked his idea—everyone, Weiser included, had laughed when Weiser suggested it—but he knew exactly where he'd go with the award once he got the keys.

Driving Highway 50 through Nevada had been a dream of his. "The desert seems to call to me at times," Weiser wrote in a letter to his mother and his sister Mona. "Its apparent simplicity, its starkness, its inhumanness. Its empty spaces suck me in."[51] He had stopped off the previous night at a Days Inn in Reno, then awoke at dawn to get an early start. The trail of asphalt before him was untrodden and discernible for miles in the distance, whereupon it seemed to vanish into the mountainside. He pulled over after a few hours to call in to a meeting at PARC; when he resumed driving, an exotic-looking patch of scattered thunderstorms took shape ahead—"the rain falling like gray veils from high individual clouds," he marveled.[52] The air was so dry that the hard rain he saw above had dissipated into a mist by the time it reached his car. He slowed his speed to 105.

Maybe this was the time to step out from the parade of lab demos and roundtable discussions and try once more to write a book. But writing didn't appear to jump-start his spirits just then, either. The grand metaphors and the boisterous, history-making tone that had animated his earlier essays did not buoy the few ideas he had published since last fall's Wearables gala. His recent prose attempted to bargain with readers, as if his natural verve had been tempered by the slight cooling of interest in his work. The paper on calm technology hadn't quite generated the industry-wide moment of reflection that he and

Brown hoped it would. Nevertheless, perhaps from reflex, the titles for Weiser's latest talks and papers began with the same three words: "The Future of . . ." His role as PARC's chief technologist still required him to issue prognostications on a regular basis—to be, as his friend Paul Saffo reportedly put it, PARC's "dancing bear."[53] At the behest of the Association for Computing Machinery, he'd just written a short piece outlining a system of credit card–like interfaces that people might keep in their purses and wallets. "Doing less can be an improvement," he insisted, hoping to stipulate his technical audiences into indulging him for a minute.[54] His essay "The Future of Ubiquitous Computing on Campus" asked readers to imagine themselves as college students having in their wallets several internet-connected cards. On the face of each card was "a bright, easy-to-read computer display with a few buttons along the edge."[55] Each computerized card performed a very limited set of related functions. The "campus map card," for example, "would always tell you where you are, and help direct you to new places. At your request, it would broadcast your location to your friend's map cards, so you could meet up."[56] Another card, which Weiser called "the food service card," stored the menus of local restaurants, notified you of any wait times, and kept a caloric record of the food you ordered every day. He alluded to a couple of other cards bearing some vague academic functions, then moved on from this card schema without mentioning it again. A more pressing agitation overtook his playful talk of new inventions.

The focus of his remaining paragraphs shifted toward questions of privacy. "The computers around us will know much more about our location and our activities than they do now," he cautioned.[57] Almost certainly on his mind here were the Media Lab's latest developments. "We must have the will," he continued, to "firmly establish a right to privacy of all personal information on any computer, no matter who owns the machine. . . . Information about me on it should be mine."[58] It was only after connecting the dots between those concluding passages with bits from Weiser's next presentation that we can fully appreciate what troubled him. It went beyond privacy as a matter of principle.

Some degree of personal information had to flow through ubicomp devices in order to render them context aware. An application like Weiser's proposed campus card map couldn't guide you to your destination if it didn't monitor your present whereabouts. Collecting personal information crossed a line, though, whenever the gathering and processing of user data were performed less in the service of the user's intentions than in the more nebulous interest of holding her attention. The brazenly humble notion of Weiser's wallet-bound, single-function computerized cards seemed calibrated to minimize the odds of sustained engagement. By contrast, state-of-the-art wearable systems like those gaining traction at the Media Lab implied the presumption of a virtuous circle anchored in constant usage: intimate streaming datasets enabled algorithms to master each user's characteristics and preferences, which in turn allowed the user to reap the fruits of the computer's labor without having to waste time inputting information and operating the software. The ideal user envisioned through this lens was one who relished being continually immersed in the digital insights presented to him by an interface agent who knew her better than she knew herself. This gift of that steady attention was the ultimate prize to be earned and won; sustained engagement proved a program's value and utility.

Weiser could see, as plainly as the desert landscape surrounding the road he sped along, that this budding imperative—to never leave the user unassisted—would drive technologists to collect all the personal information they possibly could. Furthermore, with the way things were going, it was hard to imagine this dual pursuit of data and attention leading to anything more civil than an industry-wide battle royale. The past few years on the World Wide Web, Weiser opined in a presentation with his colleague Dan Russell, had given rise to an "attention competition." Technologists, they said, were vying online to command "the scarcest resource . . . user attention."[59] Animated graphics were the weapon of the day that website designers deployed gratuitously in hopes of outshining one another. Weiser feared that handheld and embedded networked devices could intensify the competition in ways that would be difficult to control. Tech companies big and small were developing products and services without considering how each part might fit into the whole of daily life. "No single vendor should assume that their suite of devices and systems will become dominant," Weiser and Russell told an audience of computer scientists, who were mostly employed by a single vendor of one sort or another. The speakers urged the audience to instead think of user experience in the ubicomp age as "an ensemble of interactions" across many different systems that "cooperatively work together as opposed to competing for user attention."[60] Absent such cooperation, Weiser and Russell worried that engineers working in isolation would unwittingly end up building "a world where we live in a cacophony of many voices, each looking for a piece of user attention."[61] The voices with the most personal data at their disposal would speak the loudest. People stood to lose regardless of who prevailed. The sprawling, internal deserts that bloomed only in solitude were quietly being rezoned for commercial development. Maybe on some dim tacit level, Weiser's drive across America's loneliest road had been spurred by a sense of this looming prospect.

On a ten-lane expressway just outside of Sacramento, a police officer clocked Weiser's Z3 going 75 mph in a 65 mph zone, and gave him a ticket. Weiser then got back on the road, only to turn off again a few minutes later when a friend rang his cell phone in tears. A loved one of hers was dying. Friends liked talking to Weiser in dire moments; he had a special capacity to summon empathy on command, as if the sharing of pain was a higher reality that beckoned to him above everything else. They talked for an hour. "I felt good," Weiser noted after the call. "I had entered another universe for a while, and connected."[62] It was nighttime, and he realized only then that he had pulled into a Jack in the Box parking lot to take his friend's call. He got out of the car, ordered a burger, curly fries, and a large Diet Coke, and slowly drove the last hour and a half back to Palo Alto.

10

A Form of Worship

DURING THE MONTHS THAT FOLLOWED his many visits to the MIT Media Lab in 1997, Weiser found himself thinking back to an old battle he'd had with a pair of computer science professors at the start of his academic career. He recounted the dispute at length during a plenary talk he gave at a Berkeley conference on science and religion in June 1998. The two professors had tentatively rejected a journal article Weiser had written about the source code underlying a complex operating system. They noted, however, that they would gladly publish Weiser's article in the special issue they were editing if—and only if—he agreed to cover up certain portions of the images he wished to include. "Your photographs of what's going on in the system," he recalled one of them saying, "contain all kinds of things that you can't explain. Would you please white out all the parts that you don't understand and just show the parts that you do understand."[1] In addition to citing snippets of the source code he examined in his paper, Weiser also wanted readers to see a programmer's-eye view of the backstage activities that the code set into motion. The self-styled "system photographs" featured in his paper attempted to capture the torrent of commands and digits that paraded down his computer screen as it ran the OS source code. When looking over his images of these processes that otherwise passed too quickly for careful analysis, Weiser noticed interesting discrepancies between what the program's code instructed the computer to do and what it actually did, though he could not figure out what was causing the glitch.

Weiser could have implemented the professors' request in minutes, but he refused. Instead he chose to plead his case in an exchange of letters and phone calls that lasted for weeks. The photographs were essential, he told the professors, for they invited computer scientists to confront a side of their work that was unclear, yet surely worth bringing to the field's attention. Weiser insisted that another scholar might glean something from the images that he could not. The editors' recommendation to white out the photos struck him as a peculiar kind of censorship so rampant in the technical disciplines. To Weiser's mind, uncertainty necessarily enveloped any genuine act of discovery. How could any scientific field deliver on its ideals when its practitioners, in the name of rigor, distorted their work to feign a pretense of total clarity? If computer scientists shared only those findings they could fully explain, then their most perplexing observations and visceral hunches would get left out. The uncertainties that arose during the course of research and invention, Weiser insisted to the professors over and again, needed to be shared because new knowledge came from trying to grasp the unknown.

It was as if their knee-jerk condemnation of the photographs struck him as a personal affront. Indeed, Weiser's reverence for uncertainty extended beyond matters of computer science and into his basic stance toward life. There was something sacred about trusting intuition, whether it concerned a hypothesis about technology or a gut feeling about what was right. You would sometimes be wrong, of course, but the very attempt to act in the face of doubt compelled you to engage with sensations you felt but could not articulate. Intuition sprang from moments when you were still encountering things that were not yet reducible to what you had learned to make of them. A murky situation invited you—forced you—to feel your way around at the edges of routine wisdom. To habitually discard what you didn't know for certain was to dampen and dull whatever tacit attunement you might have with the world.

"Good engineering requires a good relationship to the mystery of being alive," Weiser told the audience gathered before him at Berkeley.[2]

Maintaining your reverence for phenomena that defied explicit understanding was fundamental to such a relationship. Applying white-out to conceal things you could not explain was antithetical to this mindset. During the summer of 1998, Weiser's remarks about technology took a sudden spiritual turn. The tacit dimension of human experience was, he seemed to fear, at risk of someday becoming submerged by mobile, embedded, and wearable computers that were designed to interpret situations on their users' behalf. The prospect was still quite distant, but Weiser had long been in the habit of extrapolating a bigger picture from a handful of prototypes. From the gadgets on display at the MIT Wearables gala he could foresee the onset of a wearables-laden society. The billiards player adorned in the Media Lab's stochasticks system saw one optimal shot plotted out for them the instant they looked at the table. There was no need for them to survey the table for possible shots, no need to surmise from past experience how hard they should strike the cue ball, and no need to envision the ball's trajectory. Programs like stochasticks offered up their calculations in advance of—in place of—their users' intuition. Already, members of the MIT Wearable Computing group were extending this technology to their other hobbies as a means to prove its versatility. They followed their billiards application with another designed for a homegrown game at MIT called Patrol—a contest held on weekends inside campus buildings in which students ran around shooting one another with rubber darts. The Patrol prototype, which Pentland and Starner presented at the 1998 International Symposium on Wearable Computers, installed two tiny video cameras on a hat. One camera monitored the user's field of vision and interpreted the situation before them, while the other camera watched the user's own body to identify what he was doing from moment to moment so that the software could tell him precisely what to do next. Just as stochasticks mobilized wearable cameras, computer vision, and machine-learning algorithms to determine the billiards player's best course of action, the Patrol prototype devised and delivered step-by-step instructions for locating and shooting one's opponents in the most efficient manner. Both systems aimed to assume the burden of grappling with the complexity of a complicated situation; they monitored the scene, and the user looked to his head-mounted screen for guidance.

The logic apparent in these prototypes was not too far removed from the line of thinking invoked by those two professors who asked Weiser to white out parts of his photographs. That he recognized echoes of the latter in the former was almost surely the reason he was rehashing that old spat in his talks—and the reason his remarks now centered around themes such as "the essential unknowability of the world" and "the dangers of [an] oversimplifying view of human beings."[3] In a high-tech way, the MIT wearables were primed to apply a kind of white-out onto the mysteries of everyday life. The instructions they issued to their users were generated entirely on the basis of data captured from whatever variables their sensors and cameras could measure. By attaching more and more tracking mechanisms to people's bodies, these systems ostensibly converted the phenomenological noise of fleeting sensations into a veritable database that could be objectively mined and algorithmically managed. They purported to approximate and even amplify what the expert mind did (in a game of billiards or Patrol, for starters) in order to deliver the fruits of that mental labor to anyone willing to don a head-mounted display. To Weiser's mind, this intimate coupling of body and interface threatened to weaken our sense of connection to the world.

Calculation and intuition, Weiser surmised, were like two beacons disparately orienting the diverging R&D initiatives that were now pulling apart his original vision for ubicomp. His initial notion of a room filled with a hundred computers had proven itself a highly variable proposition. One could imagine a hallway where the rooms and people on one side were outfitted with smart devices built by Negroponte's butler-oriented protégées, while the rooms on the other side featured calm technologies designed by Weiser, John Seely Brown, Rich Gold, Natalie Jeremijenko, and Hiroshi Ishii. The smart rooms on the former side would be strewn with gadgets collecting the most data possible to optimize the AI systems that oversaw those spaces and monitored their inhabitants. The other rooms, on the side crafted by Weiser and his collaborators, would make no such attempts at technocratic control. The computerized objects in there would convey, imply, or hint at meanings for people to sense, interpret, and decide to act upon or coolly ignore. Their subtle machines would run on limited input, with their output intentionally confined to the power of suggestion. In those rooms, amid the digital cues coloring the periphery, there would remain a space—a need, even—for humans to fumble through the uncoded voids.

In the spirit of preservation, Weiser devoted the last paper he'd end up publishing to a new concept he termed "spiritual engineering." The paper's first sentence bore traces of the reflective, retrospective mood that had gripped him through his recent travels, from his solo drive across Nevada to a *BusinessWeek* workshop in Bali, where he stayed days after his presentation to witness traditional performances of the Barong dance in Hindu villages lining the hilly green inlets along Denpasar's eastern coast. Between his usual conference appearances and regular meetings at Xerox's Rochester headquarters, he made sure to insert two summertime trips with his older daughter, Nicole. They went to New Jersey to see Weiser's younger sister, Mona, who had recently been diagnosed with cancer and with whom he had kept in close touch since he left for college. Mark and Nicole also flew to Greece and walked among the ancient ruins. They lingered in silence before the shrine of the Oracle of Delphi, where the Greeks would trek to pose their gravest questions about the future. They toured the rubble of the agora of Athens, where Socrates waged his famous arguments. Around the marble columns of the Parthenon, father and daughter photographed each other in the city's white morning light as each contemplated its history. Weiser strode slowly around the Acropolis in his bouncing gait, smiling and in tears. "I felt in heaven," he noted of the excursion, "visiting these old sites of my philosophical readings."[4] The joy he had seemed to lose hold of over the last few years—the palpable glow that had warmed his eyes at the sight of an idea—flared up among the crunch and dust of those fallen remains. He returned to his office in Palo Alto ready once again to issue an unusual challenge to his fellow technologists.

"I have been doing engineering for more than 30 years—in start up companies, in big companies, as a student, as a professor, for a hobby, for an escape, to lose myself, to find myself," wrote Weiser in the opening of his final essay.[5] He described how he had lately come to regard engineering as "a form of worship" not so different than religious prayer. Both practices, as he saw them, involved "honoring the unknowable connection among things."[6] He conceded that most engineers found this comparison ludicrous whenever he aired it in conversation. Whether they constructed bridges or software, engineers took pride in their precision. It was the duty of their profession to make every necessary calculation to ensure structural integrity, operationalize a system, and fix any bugs. Weiser prided himself on all of this, too. But his conception of engineering went far above materials, physical laws, or computer code. He had hit a point in his career where the debates he was having with other technologists about their newest inventions all seemed to be tangled up in a bigger question—a question that stepped back from the gadgetry to interrogate how these engineers of the digital world collectively viewed themselves, and how they viewed other people as well.

Weiser's personal mission as a computer engineer, he wrote, had been "to fit technology to humans."[7] He had deployed this phrase in the concluding sentence to his *Scientific American* article and mentioned it in many talks, but it wasn't enough to signify all that he meant. About *humans* he should have elaborated further. The problem wasn't that most technologists lacked interest in making devices that fit seamlessly into people's lives. The problem was that technologists generally carried with them a mechanistic, oversimplified picture of the species. They tended to think of themselves this way, too, at least insofar as their work habits let on. It was an updated version of the same objection Weiser had described in the audio letter he made in Ann Arbor to send to his father in Stony Brook: "We understand that there is to the world a mystery, an ineffableness. . . . And this is not intolerable to us, as it is to some people, who must fight against this, expose it, strip it bear."[8] One way to make sense of Weiser's recent emphasis on spirituality in his talks on technology was to understand it as a stand he was taking on behalf of the values that bound him to his father. "It's letting go into the mystery that is the most human activity," Weiser's letter had continued. "You and I know this."[9] And so, setting ubicomp aside, Weiser devoted the bulk of his present essay to expounding upon "the fullest sense of humanness" that he urged more engineers to adopt—a sense that, he insisted (in a jab at the MIT Wearables group), "extends well beyond what is inside our skins or our heads."[10]

Readers familiar with Weiser's previous writing could guess which ideas he'd draw upon here to liven up his own. Entanglement, attunement, tacit knowledge, center/periphery, the white cane example—all would be stirred into the mix again. Here, though, unlike in his earlier pieces, Weiser layered these concepts to form a theory about the process of creating technical objects. He had applied this line of thinking almost entirely to technology users before, as if they were the only side in the innovation equation whose mentalities needed to be illuminated. Engineers, he now felt compelled to say, were not exempted from the insights of Polanyi, Heidegger, Harding, Merleau-Ponty, and the rest. Everyone was an iceberg: clear and discrete on the surface, murky and porous below. "At any moment of our being in the world[,] much is hidden to us," Weiser wrote.[11] The submerged majority included "the world of our mind."[12] The scope of an engineer's consciousness was fundamentally limited, more or less like the rest of us, by the bodily confines that engendered her attention. In order to concentrate on one aspect of a specific thing—be it a dataset or a speech—we were perpetually letting something else fall out of our focus. Amassing more information offered up more things to successively direct our focus toward, but it did not alter the partial nature of attention. "Unknowability," Weiser asserted, was "an essential fact of being human."[13] No amount of sensors, screens, or intelligent assistants would change that. Fortunately, our capacity to act, build, and understand was not restricted to what we knew explicitly. "There is always much more unconscious than conscious in our behaviors," Weiser continued, as he pivoted to emphasize a paradoxical takeaway he wanted engineers to gather from the philosophy.[14]

On one hand, this iceberg model of the mind appeared to invite a diminutive, skeptical view toward the whole engineering enterprise. Weiser assured readers that his was not a nihilistic position. Objects and systems still needed to be constructed in spite of the uncertainties that surrounded them. But an attitude adjustment on the part of engineers was long overdue. Weiser suggested that engineers were either lying to themselves or being willfully reductive whenever they clung to mindsets and models in which every variable of their work was thoroughly accounted for. If you were sure that your invention would be for others exactly what it was in your blueprint, then you were thinking only about the visible tip of the iceberg. You were presuming to know more than you did—your knowledge of other people was always incomplete. Even *their* own self-awareness was largely tacit, as was your own. Better to acknowledge these uncertainties, Weiser advised, than to presume yourself into an illusory state of omniscience. Adhering so doggedly to the pretense of certainty could push you to dismiss variables beyond your control or, worse, warp them to serve your design. Instead, Weiser urged his fellow engineers and technologists to cultivate an attitude of "deep humility," which ought chiefly to encourage "humility toward the role of [our] artifacts in other people's lives."[15] In Weiser's estimation, a fabricated invention entered into society "like a prayer," made by an engineer, "in relationship to what can never be fully known."[16] Technologies should thus be regarded as speculative offerings and not miracles bestowed from above, for innovators were no masters of the universe.

And yet, a humble assessment of engineering could be empowering in its own way. The uncertainties that encompassed any project gave the builder of new objects a poetic license, Weiser argued, to take modest leaps of faith. Granted, neither science nor engineering would benefit from the return of superstition or sorcery. But reactions against the more errant forms of subjectivity went too far whenever they rationalized turning a blind eye. When technologists "design objects only for our conscious attention"—when they habitually rejected notions deemed soft or fuzzy for want of harder data—"they miss[ed] the biggest part of us," wrote Weiser.[17] "Objects so designed [would] be clumsy, and [would] not 'fit' us."[18] By conscientiously ridding their creative process of subjective impurities, technologists engineered devices that tended to underplay their intended users in a corresponding manner. A device forged without intuition failed to feel intuitive because its design fails to imagine how people might tacitly attune to it in practice. The leap of faith was not taken, likely not even discussed, owing to the occupational reflex to steer clear of murky terrain.

By contrast, Weiser suggested, engineering practiced as a form of worship would strive to provide "objectification of what is mysterious about being human."[19] Built environments and digital technologies should include subtle features that stem from and gesture toward our unconscious drives and desires. The best example of this from the personal computing paradigm was Apple's original Macintosh monitor: "The case reminds us of a face," Weiser noted, as elements such as its floppy-disk drive evoked a smile "without being too strong."[20] A world laced with smart devices made it all the more imperative to build ubicomp objects that offered more than the sum of their functionality. Without such efforts, we would be surrounded by instrumental entities that all connected to one another but left us feeling alienated. Our wearable technologies would guide us through life with a steady stream of metrics and instructions, displayed in our glasses or spoken into our ears, which we could follow without having to make sense of the situation ourselves. Weiser's brand of spiritual engineering postulated that inventors had a social responsibility to "act beyond the data . . . on probabilities and possibilities and hunches."[21] Technologists needed to value intuition as much as they valued calculation; failing to do so led them to create systems that left no room for intuition to emerge. Intuition bubbled up from the sensory grasp our bodies maintained with forces at the periphery of consciousness. Intuition fashioned scaffolding from what was said so that we might sense the unspoken. Intuition was, in this sense, the very root of connection. To disregard our intuition in deference to streaming information cut us off from that deeper, dimmer communion with the world, and so promised to diminish the expansive humanity Weiser sought to design toward.

The essay ended without mention of any new technologies Weiser had on his mind. In his previous writings he always highlighted some prototype as a harbinger of promising developments. Whether or not Weiser's critique of mechanistic reasoning was tied to a budding ambition he would pursue in his lab remained unclear. For the first time in a long time, his notebooks bore no doodles or phrases that might spur a next phase in his ubicomp quest. Whatever plans he might have had incubating below the surface were replaced by a new thought in March 1999.

At Dan Russell's urging, Weiser eventually went to see a doctor. Russell had recently returned to PARC after working at Apple for several years. The research team Russell led at Apple had been developing a tablet computer, and Steve Jobs had decided that tablets were a worthless proposition. "Nobody would want a tablet computer," Russell recalled Jobs saying as the CEO ordered him out of his office.[22] Brown seized the chance to hire Russell's entire team and have them continue their project under Weiser's management. Shortly into the new millennium, they hoped, PARC might perhaps have a market-ready tablet—with wireless connectivity, a camera, and a multitouch screen—that could become a standout product for Xerox. While tablets no longer epitomized Weiser's long-term vision, he still believed they were an important stepping-stone whose initial design would set a tone for the next decade. The tiny computers of the 1990s, such as the Poqet PC, the Newton, and the PalmPilot, each functioned like miniaturized desktops. A new tablet designed at PARC and sold by Xerox might bring a bona fide ubicomp device—its "phase I" features, at least—to the masses for the first time.

Russell was one of the few colleagues who had seen Weiser regularly throughout 1998. All year, Weiser had been flying between various research labs, conferences, Xerox facilities, and international trips. A week divided among three different cities was not uncommon. Even around town he seemed to be in constant motion. Sometimes he would hurry from his return flight home and take Corinne to his band's jam sessions, where his daughter would sing along as he played the drums; other times he would zip from the airport to a local airfield to go sky-diving—a new hobby he had added for the sensory rush (and Corinne joined him for that, too).

Whenever Weiser was in the office, he met with Russell frequently to discuss the tablet project. For months Weiser had been dealing with a pain in his stomach that had spread to his chest by March 1999. It had grown from an irritant Weiser rarely mentioned into an agonized expression that chronically robbed his face of its usual smile. "Just go talk to your doctor," urged Russell.[23] Weiser had been seeing his psychologist as often as possible to talk about his childhood, but he had not had his body examined at all since the pain began.

Later that week, Weiser came into work carrying the slide of an X-ray he had just gotten at a nearby hospital. He was taking it in to have it diagnosed by his radiologist that afternoon. Weiser didn't know how to read an X-ray, but Russell did. He asked Weiser to hold the slide up against his office window. "No, Mark, turn it over," Russell instructed when Weiser held it upside down.[24] After Russell stared at the X-ray for a minute, he simply turned to Weiser and asked him to call him after the appointment.

Weiser didn't know that Russell had been trained to identify cancer in X-rays for a medical software project he had written code for back in graduate school. The moment Russell saw Weiser's X-ray, he knew that it was a terminal case. "It was, I don't know, fifty tumors on his chest and stomach and so on," he said.[25] But at that moment, he could not bear to tell Weiser what he saw. Anyway, it seemed best to wait for an expert opinion.

When Weiser called from the radiologist's office, he confirmed Russell's worst suspicions. "I have stomach cancer," Weiser said to Russell. "Three months to live."[26]

The situation was spiraling so fast. Soon after Weiser told Vicky Reich about the gut-wrenching diagnosis, he was telling her how he intended to spend his remaining days. They had not spoken very much since Weiser moved out of the house.[27] Once Reich learned of his condition, she and a few of his friends took turns driving him around town to seek some alternative treatments he wanted to explore. None promised much help at such a late stage. In lieu of hope, Weiser threw himself desperately into making a decision about his next and final move.

He needed to come up with a vision confined to the stark, untimely present. To what should he devote the little time he had left? To whom? He thought he had found the answer in teary chats with his PARC colleagues. His plan was not entirely surprising, but it shocked Reich to hear him say it, and to see that he meant it.

"I want to write the book I never got the chance to write," Weiser had declared in a conversation with Brown.[28]

And he wished to write it by the sea. For days he was fixated on securing a beachfront property in Half Moon Bay.[29] Any place he could move into quickly, with a view of the ocean, where he could look out at the waves breaking in between sentences. He longed to think deeply and quietly along the coast again. This was the same reflex that had compelled him to spend afternoons walking the shorelines of his Long Island youth. The same yearning for mystical ideas that kept him reading unknowable paragraphs at New College as dusk fell around the library and Sarasota Bay, its green shallows lapping against a vacant dock. The Pacific beckoned like memories of a distant home. He imagined himself lying on a recliner before a west-facing window, typing on a special laptop PARC would rig up for him, trying to string together words that might get technology back on track toward the future he envisioned but would not see.

The irony of this scenario could not be lost on him—that his dying days would be passed in relative isolation, staring at a PC, what he called nine months earlier "a life-distorting device."[30] But this book he wished to write would not dwell on the limitations of PCs. That critique driving his original arguments had become a familiar talking point recited by the computer scientists who now worked on mobile, wearable, or embedded systems. The post-PC revolution was already underway in the world's top R&D labs.

Most concerning to Weiser were the multiple paths this revolution's development might follow. There was the path that the MIT Wearable Computing group was rapidly forging, which could lead to a sensor-filled society where people wore glasses that told them everything they wanted and needed to know. There was also the way of MIT's Things That Think consortium, whose more discreet approach to computerizing things better approximated the philosophy of calm technology that Weiser and Brown had begun to articulate. Still, the TTT prototypes demonstrated a greater enthusiasm for automation than Weiser endorsed. It was a slippery slope to prioritize a system's capacity for automation over an individual's sense of agency—that is, to chase superior outcomes without grave concern for how involved people were or weren't in the process. Future TTT-style products could be empowering if they were carefully designed to be, or they could slip in the technocratic direction of Jim Clark's omniscient, ominous smart boat.

Perhaps in time Weiser could have guided the technology's progress little by little, the way a parent tried to raise a child. His only recourse now was to set down in advance everything he might ever say. To type into his laptop all the questions he might ask his colleagues in a brainstorming session, all the little anecdotes he might share with fellow researchers over beers and laughter, all the words of caution he might impart to a public that had no real idea about what was coming.

He needed his book to be a map showing others how to carry on the twenty-year quest he had plotted out in his thirties, twelve years before. He wanted to clarify "the real essence of ubiquitous computing," Weiser told Brown. "They've completely missed the non-technical part of what ubiquitous computing is all about."[31] In addition to illuminating for technologists the philosophical underpinnings of his vision,

Weiser wished to alert lay readers to the excesses and abuses he saw brewing in cutting-edge prototypes that so many journalists were dutifully hyping, as they had often done for his. He hoped his book would be "something of a bill of rights for people to hold up against the coming wave of smaller, smarter computers."[32] The notion of a hundred computers per room had acquired, with the proliferation of its emerging varieties, the contours of a prospective empire.

As far as Reich could tell, she was the only one urging Weiser to stay.[33] If writing his book meant living alone on the coast, then that meant he wouldn't be seeing their daughters as much; the wooded road that wound through the Santa Cruz Mountains, from Palo Alto to Half Moon Bay and back, was not a crosstown drive they could make after school. It was too far for his PARC colleagues to simply drop by after work. As Weiser ventured to take one last pass at the details of his vision, his family and friends would be left most days thinking about him from afar. They could send emails, of course, or call him on the phone. They could probably set him up with a system to handle video calls, and perhaps carry PARC tabs to let him know their whereabouts. They could use the labs' gadgets to stay connected in every way the internet made possible.

But the ever-expansive reach of digital technologies was, at least to Reich, hardly something to marvel over in that moment. The thought of Weiser not being wholly there, while he was still able to be, felt like "the worst scenario anyone could dream of."[34] Reich persisted amid the chorus of researchers who were seconding his notion to write in seclusion. "No, Mark," she told him repeatedly. "You have to be home."[35]

The crisis forced Weiser to confront a question that had always hung above his work: Which kind of connections mattered the most? He had been dreaming up a world where one did not need to choose between physical and digital, online or off, sensation versus information. He believed every object, every room, held an opportunity to give people the best of both. Stomach cancer forced him to choose. He could hammer out a book in a beach house by the sea, or he could return to the comfort of his family's home. Weiser agreed to move back.

In one sense, as it turned out, this decision didn't seem to matter very much. His condition worsened quicker than anyone expected—there would've been no book either way. In another sense, his final move could be counted among his most profound.

With Weiser at home, Nicole left college to come home as well. Reich and their daughters swiftly rearranged the house to make space for Weiser's hospital bed in a big open room by the kitchen where everybody liked to hang out. They scooted his bed beside the sliding-glass door so he could look outside and friends could visit him easily. They took him for walks around the neighborhood, then just along the block. Eating was hard, and it soon became unbearable. Within three weeks of his diagnosis, he began to look noticeably thinner by the day. His sister Mona, who had herself been battling cancer for more than a year, had flown out to see him when he'd gotten the news. Shortly after she flew back east, he had to say goodbye to her over the phone from his hospital bed as she passed away on hers.

Pain marred the last few weeks of Weiser's life. But this time was not without joy and revelation. For many years he had not stayed in one place this long. The dizzying list of canceled speaking engagements showed just how hurried his routine had become. Around him the house was still. He watched as others came and went while he remained there on his bed. Gone were the orange extension cords that had, for a summer, connected the tabs and pad he brought home for the family to play with. None of the stuff now in his midst boasted an IP address, aside from the new laptop and dictation system that PARC had promised and delivered. Weiser opened neither.

His parting visions were about his family's future. Those waterfront hours of writing he had pictured were filled instead with conversations. Together they smiled over forgotten memories. Everyone had their own favorite image of Weiser. Like the day he had entered Reich's life, when they were teenagers running around the Museum of Natural History. Or the time he emerged carrying a stack of physics books after his youngest daughter, Corinne, then six, had asked her father why the sky was blue—and she was treated to his most exuberant forty-five-minute explanation. There were also apologies, reassurances, and regrets. Reich and Weiser spoke about the people she could rely on after he was gone. It moved her to see him straining against the pain, striving to help in his waning thoughts.[36]

Weiser's eyes had always been expressive windows to his mind. Even during his final days, standing by his bedside, visiting friends noticed him mulling over some inaudible notion. This house was so different than the one he grew up in. Gathered all around him now was evidence of bonds he had managed to build. His stacks of books included those he had read with Nicole, in between his travels and during adventures he took her on. They had chosen the books together, discussed them together, furrowed their brows over them together. The Xerox-branded notebooks and pens around the house belonged to his daughters, who had brought these trinkets home after spending a day at PARC, as they so often did at their father's invitation. Nicole and Corinne knew the colleagues who kept dropping by to see him—knew them far better than he had known any of his father's colleagues, because Weiser had made it a point to include his daughters in conversations, even at lunch tables and conference rooms filled with his fellow PhDs.[37]

And lining Corinne's bedroom were the dolls her father had brought back for her after every international work trip. Each doll was from a different country, where she was a constant in his mind. Like Nicole, Corinne would remember the many computers he had brought home for them all to tinker with over the years—both daughters would pursue careers in tech. But it didn't matter that the dolls were just dolls, without microchips or special sensing abilities. Corinne felt in them the presence of so much she had come to love about her father, and she would always keep them on display.[38]

It must have hurt Weiser to think about how unfinished his life's work was. It must have also occurred to him then, during one of his medicated gazes at the people and things in the big room surrounding his bed, just how genuinely connected he was, even without a ubicomp device in sight. His deepest affinities would endure in the simplest things. The people in his life wouldn't need smart objects to remind them of what he meant to them. For his collaborators, a stroll around the hallways at PARC would be enough to channel his input on their future projects. For his daughters, the sight of a dusty physics book, or a long look up at the cloudless sky, could actuate the sound of his ecstatic voice raving about the atmosphere. A Polaroid taken in Athens could transmit an undying embrace.

Rich Gold, Weiser's ubicomp collaborator, stood among the family and friends gathered around Weiser's bedside on the clear, windy evening of April 27. Weiser spoke his last words to them. "Have a ball," he managed to say faintly, utterly sincere.[39] Then they watched him fade. As he did, Gold saw for a second the familiar wrinkles of Weiser's happiest expression. "When he died there was a warm and whimsical smile on his face," Gold noted, "as if he just had another brilliant idea."[40]

EPILOGUE

TALK OF UBIQUITOUS COMPUTING quieted after Mark Weiser's passing in 1999, though it would remain an area of emphasis at Xerox PARC for another decade. The center itself, however, began to change with the new millennium. As it did, Weiser's closest colleagues left and largely went their separate ways.

John Seely Brown retired from his role as PARC director in June 2000, following the publication of his book *The Social Life of Information*. He embarked on a second career as an author, speaker, and innovation consultant, while serving on many boards and collecting nine honorary degrees. Later in 2000, Rich Gold was laid off along with the entire research group he had only recently assembled. Gold's group—called RED, short for Research in Experimental Documents—was perhaps the last and loudest expression of the freewheeling, business-agnostic innovation that only Xerox PARC researchers were allowed to pursue. "If PARC was Xerox's ponytail, we were PARC's," said the ponytailed Gold.[1] The group had spent eighteen months creating wildly futurist reading devices and readying them for public display at the San Jose Tech Museum of Innovation. Their exhibit was seen by four million people and featured on *ABC World News Tonight*, and would be remembered as a formidable example of speculative design. The group's firing sent a message that even the center's most resolutely technical scientists found alarming and sad. "Xerox could no longer afford to support this level of play," noted Eric Saund, who had joined in 1988 and would stay until 2017. That was the moment, he recalled, when "our childhoods ended."[2] After leaving PARC, Gold began writing a book that he nearly finished before he, like Weiser, died young after battling cancer. Published posthumously, *The Plenitude* (2007)—Gold's autobiographical, cartoon-filled account of the lessons he learned through a life making art, music, toys, and technologies—was edited and introduced by his wife and artistic collaborator, Marina LaPalma.

The Work Practice and Technology group, Lucy Suchman's team of anthropologists and ethnographically inclined technologists, disbanded as well. Suchman had always been a shrewd observer of corporate cultures. She sensed earlier than most how the ongoing campaign to reinvent the center during the late 1990s—an initiative dubbed "PARC 2000"—spelled the end of an era. These all-staff strategic planning exercises, Suchman recalled, gave rise to "this growing narrative around the idea that PARC had run out of steam."[3] No one on her team subscribed to that view. At the time, they were in the middle of a major project with civil engineers at Caltrans who were designing a new bridge to span the San Francisco Bay. Suchman's team, in the spirit of their previous work with the Silicon Valley law firm and the San Jose International Airport, studied the civil engineers' workflow while building them a tailor-made electronic file repository to support project documentation and online collaboration. Her team's efforts over the past decade had produced scores of academic publications that stood at the forefront of multiple fields. And yet, given the tensions mounting at PARC, they suddenly found themselves in a double bind. "Rather than [seen as] responsible action, our reluctance to abandon existing projects . . . was read as a kind of recalcitrance, a form of resistance to change," Suchman later wrote.[4] The prospect of falling in line with the new agenda seemed equally desolate. There was an internal push to become "startup-y" and to make "the next billion-dollar app," Suchman said. "There was just no fit between the kind of long-term research that we wanted to do and that."[5] And so, feeling like they were being forced out, the group spent their remaining budget on "a big intellectual party" in June 1999.[6] Officially, the event was called Work Practice & Technology: The Next 20 Years of Research. Scholars from around the world came to gather in PARC's auditorium. "It was a time to celebrate with wide network of friends and fans and people who we had learned from and loved and appreciated what we had done," said Suchman.[7] She is now an esteemed professor at the University of Lancaster. Not long after she cleared out her office at PARC, Suchman was awarded the prestigious Benjamin Franklin Medal in Computer and Cognitive Science.

In January 2002, Xerox decided to alter its relationship with the Palo Alto facility, spinning it off as an independent subsidiary. Xerox PARC became "PARC Inc., a Xerox Company." This new name was the last step in a series of adjustments that made the center more like all the other technology ventures in Silicon Valley. As the decline of Xerox funding sank to new lows, PARC scientists were encouraged to take on pragmatic assignments that would help pay for their research. They needed to think of Xerox less as a familial benefactor and more like a client among many whose business had to be earned. In practice, this forced the center to realign its agenda around tasks the staff had previously waved off in favor of pure science. The old attitude toward intellectual property was a case in point. For decades, Xerox PARC had housed brilliant technologists who openly despised patents on the grounds that their ideas should be available for others to build upon.[8] This scholarly ethos proved a costly virtue when it allowed upstarts like Google to use PARC's advances for free.[9] Patents and licensing would be front and center for PARC *Inc.*, as would the pursuit of government funding and "strategic research engagements with other companies."[10] Urgent as these priorities were from a financial standpoint, the new direction promised to turn PARC into a place that many of its seasoned researchers wanted nothing to do with. Some returned to academia, where they could remain insulated from the Wall Street incentives that had swiftly remade Silicon Valley into a quarterly minded casino. Others decided to play the game by joining a better team, and they had their pick of offers from Microsoft, Google, IBM, or Intel. Incidentally, these sweeping changes to Silicon Valley's business environment strengthened the MIT Media Lab. While declining corporations like Xerox scaled back on basic research, younger dot-com firms harbored little interest in creating their own PARC-like centers for blue-sky R&D. Instead, technology companies mostly looked outside themselves to glimpse radical inventions being prototyped elsewhere—and the Media Lab was often the first place they went.

The dissolution of Xerox PARC was primed to happen regardless of its chief technologist. Weiser's sudden death had little to no bearing on Xerox's sprawling corporate interests. But the center's transformation, which began in earnest months after Weiser passed, made for a rather dispersed uptake of his unfinished quest. His young recruits who were perhaps best positioned to bring ubicomp into the twenty-first century each relocated to run university labs at different corners of the US. They remained committed to the vision and saw one another at conferences, but never again worked in the same building.

On a global scale, however, computer scientists came together to commemorate Weiser's legacy in ways that extended his influence on research for decades. A prestigious journal—*IEEE Pervasive Computing*—launched in 2002 with a mission, which it still bears today, to "act as a catalyst for realizing the ideas described by Mark Weiser."[11] The journal's founding editor, the Carnegie Mellon professor Mahadev Satyanarayanan, dedicated the first issue to Weiser as he noted, bittersweetly, how the technological infrastructure was finally catching up: "After a decade of hardware progress, many critical elements that were exotic in 1991 are now viable commercial products. . . . We are now in a better position to begin the quest for Weiser's vision."[12] An international conference series, too—ACM's annual UbiComp conference—quickly grew up around Weiser's writings. "Almost one quarter of papers published in the UbiComp conferences between 2001 and 2005 cite Weiser's foundational articles," reported the anthropologist Genevieve Bell and computer scientist Paul Dourish, "a remarkable number of publications to cite a single vision as fundamental for their own work over a decade later."[13] Dourish, who had worked with Weiser at PARC, wrote an influential 2001 book that emphasized the philosophical underpinnings associated with post-desktop interfaces, effectively continuing and extending conversations with phenomenological thinkers that Weiser had initiated.[14] At Intel, Weiser's close friend and collaborator Roy Want led considerable advances in mobile hardware and wireless networking, demonstrating how newer forms of connectivity, such as RFID tags and Bluetooth, could support a multitude of context-aware applications reminiscent of the PARC tab and its attendant network of infrared transceivers.

While the researchers who had either worked with or been inspired by Weiser were keen to weave the internet into the social fabric of everyday settings, the technology companies leading the personal computing market were understandably content to build stand-alone devices that functioned like miniature PCs. The most famous and influential of these was the iPhone, which Steve Jobs unveiled on January 9, 2007. The iPhone, Jobs told his Macworld audience, gave you "the Internet in your pocket."[15] Other "smartphones" then on the market did so as well; the BlackBerry phones and the Motorola Q featured a web browser, email, and GPS. But what truly separated Apple's new iPhone from the competition, Jobs insisted, was its "desktop-class applications . . . not the crippled stuff you find on most phones."[16] Apple had done a far better job of bringing PC-quality graphics, functionality, and software to the mobile screen. Jobs underscored with great pride how closely surfing the web on iPhone's Safari browser resembled surfing the web on a laptop or desktop.

The iPhone's staggering takeover of the smartphone market quickly set a gold standard that Samsung and Microsoft would later adopt. Wi-Fi-enabled tablets, too, were designed and advertised as a product that blended desktop-level capability with mobile convenience. While Apple's integration of Google Maps gave its users a powerful way to understand and navigate their surroundings, almost every other popular mobile application kept the user's attention fixated on the screen. For much of his on-stage iPhone demonstration Jobs was, inevitably, staring down at its 3.5-inch display, providing a less intended sneak peek at how smartphones would require us to attend to them constantly in ways that Weiser dreaded. Six years prior to the iPhone's launch, channeling Weiser's sentiment, Dourish had argued that transposing PC-era design conventions onto ubicomp systems "would be a terrible idea in a computer that I'm using while driving, or crossing the street, or trying to enjoy a conversation."[17] This hypothesis would, as smartphones proliferated, be validated time and again by statistics indicating the dangers of distracted driving, by memes and videos showing iPhone-touting pedestrians bumping into stationary objects, and by stacks of books, such as the social psychologist Sherry Turkle's *Alone Together* (2011), that documented the weakening of interpersonal relationships as more people turned more often to their phones, even while gathering with family and friends.

By the time of the iPhone X's release in 2017, the operating system had evolved to become more context aware as it gradually incorporated technical advances made by the ubicomp research community. In the tradition of PARC's tab, newer iPhones could launch apps, offer reminders, and play media that coincided with the activities one engaged in at a particular location. "For example," a tech blog rejoiced, "when you plug in your phone at the gym, it automatically loads up an energetic jams playlist."[18] At present, ongoing progress in wearable computing, 5G networks, and augmented reality have Apple and other tech firms making big investments in computerized eyewear—a form factor that could spur a cleaner break from the PC era.

A digital future premised upon wearable interfaces and the Internet of Things has, of course, already arrived in some cities, workplaces, and quite a few homes. As these technologies continue to develop and spread, the engineers who design them and the corporations overseeing them will face tough decisions about where to stand on the spectrum of visions that took root at Xerox PARC and at the MIT Media Lab in the 1990s. Whether they recognize it or not, today's post-PC innovators are standing on the shoulders of either Weiser and his collaborators or Negroponte and his protégés. The ones who lean toward the latter have taken an early lead in the race to make everything smart.

Shoshana Zuboff has, at length, traced the influence that Weiser's contemporaries in the Media Lab came to exude within the biggest tech companies of the twenty-first century. She characterized the computer scientist Alex Pentland's standing in 2019 as "something of a high priest among an exclusive group of priests" whose research "is showcased in surveillance capitalist enclaves," where many of his former students now work.[19] The "reality mining" techniques he pioneered—which deploy many sensors to track everyday behaviors at a social scale—effectively demonstrated, for Google and others, how collections of personal data gathered from mobile and wearable devices could be lucrative, as they provided a basis for accurately predicting (and ultimately swaying) users' actions. Interface agents often play starring roles in current mobile, wearable, and smart-home gadgets. The latest expressions of Negroponte's butler metaphor, Google Assistant and Amazon's Alexa give voice to each company's smart speakers, all while parsing our voices into biometric data that might reveal our emotional states, personality traits, and demographic profiles.[20] Affective computing, too, is becoming a valuable means of data extraction for consumer espionage. Rosalind Picard, who initiated this field at MIT, has to her credit resisted the advertising industry's uptake of her research—so much so that she was "pushed out" of Affectiva, a startup she cofounded in 2009 to build emotion-recognition systems designed to help children with autism boost their emotional intelligence.[21] Affectiva now provides its technology to "seventy percent of the world's largest advertisers," and, in 2021, the company merged with its European competitor called Smart Eye to create "a transatlantic AI juggernaut."[22] Facebook has been patenting comparable software of its own since 2017, which could enable the company to monitor users' affective responses to posted content via the camera in their smartphones.[23]

Well intended or not, technologies that couple big data with proprietary algorithms stand to worsen the growing mismatch between technocrats who purportedly hold a "God's-eye view of humanity" and everybody else—the citizens and consumers who do not. Seemingly progressive calls for data ownership and ethical AI fail to address the extreme power-knowledge imbalances at play. Pentland has championed a "New Deal on Data" that would "give consumers a stake in the new data economy."[24] While the idea of receiving monthly kickbacks from Big Tech in exchange for our personal data has obvious appeal, granting everyone a vested interest in surveillance capitalism would only deepen our societal attachment to systems that reduce most people to numerical inputs and subjugate us to algorithmic management—administered in the form of notifications, nudges, social credit systems, or physical punishments, depending on who commands technocratic power over a given population. Meanwhile, the staggering amount of natural and financial resources needed to run large-scale AI systems render them not a little oligarchic. To the extent we entrust AI to guide decision-making and problem-solving, we risk disavowing the democratic process. We of course need regulations to curb the worse abuses of emerging technologies, but we also need these technologies to be designed differently.

Weiser's life and work, his unfinished quest, point toward a better Internet of Things—a better digital future—that can still be built. Though he would've been appalled by surveillance capitalism and by China's "perfect police state" in Xinjiang, none of it would have surprised him. "Although he did not name it," Zuboff writes, "the visionary of ubiquitous computing, Mark Weiser, foresaw the immensity of instrumentation power as a totalizing societal project."[25] (And it was Suchman, Gold, and Jeremijenko who each pushed Weiser to see that.) At present, these dystopic dimensions have grown utterly apparent, and the desire for alternative visions is spreading as a result. The excesses of Big Tech's drive to capture more data in order to more fully command users' attention have become a political issue of global concern. Silicon Valley insiders are speaking out against the industry, and their recent criticisms echo principles that Weiser preached throughout the 1990s. The ex-Google ethicist Tristan Harris and Facebook whistleblower Frances Haugen, for instance, each appeared on *60 Minutes* to expose how their former employers seek to maximize screen addiction and, as a means of doing so, willfully amplify content that is most likely to enrage their users. In place of anger and dependency, Harris's Center for Humane Technology urges his fellow designers to create technologies that foster mindfulness and "strengthen existing brilliance . . . without taking over people's lives."[26] Striking this balance, as Weiser struggled so passionately to do, will not come quick or easy. But Weiser and his collaborators have left today's like-minded innovators with a trailblazing array of concepts, prototypes, lessons, dilemmas, and challenges that help show the way forward.

Weiser believed that tiny computers and the boundless internet they wrought could evolve to play a modest supporting role at the periphery of our lives. Ubiquitous computing was never premised upon the inherent value of hardware, software, and online information; rather, it was his proposal for how digital technologies might be remade to become worthy components of the world that preceded them.

Computation, as he saw it, can be "part of the informing environment, like weather."[27] Familiar objects can be digitally inflected to communicate with us in a more tacit manner than flashy screens do. The Internet of Things can be an opportunity to infuse our everyday settings with timely, sensory expressions of data optimized for public reference. Smart cities can be made to prioritize individual citizens and local communities; they can be designed, above all else, to "increase [our] ability for informed action," like Weiser insisted.[28] In the same breath, as we scale up from Weiser's "one hundred computers per room" to a planet of one hundred billion computerized things, we would do well to remember that even the "father of ubiquitous computing" was quick to acknowledge the limits of computing. Artificial intelligence shortchanges our embodied intuition. Virtual reality—the metaverse—excludes "the infinite richness of the universe." And more often than not, feeling connected has little to do with digital connectivity.

ACKNOWLEDGMENTS

I NEVER MET MARK WEISER. When he passed away, I was just a middle schooler trying to pass my typing class. About ten years later, in graduate school, I happened across his essay "The Computer for the 21st Century" while hunting for something else in the library. I read it. I read it again more slowly. Then, ten years after that, I found myself reading nearly everything he had written and all that he had saved. Through the waves of the pandemic I wrote about his life.

That process and this book were made possible by the Special Collections staff at Stanford University Libraries, where Weiser's papers are housed. Without their efforts to catalog and organize such an extensive set of materials, my research would've taken twice as long. Thank you as well to Henry Lowood, the Harold C. Hohback Curator for History of Science & Technology Collections at Stanford, and to Leslie Berlin, project historian for the Silicon Valley Archives. My deepest thanks to Vicky Reich for gifting Weiser's papers to Stanford and making them accessible. Thank you, Vicky, Nicole, and Corinne, for sharing your memories with me. Having your family's support made my work all the more meaningful.

My archival research was greatly enriched by conversations with Weiser's friends and colleagues who worked at Xerox PARC. Thank you so much to everyone who granted me an interview, many of which were conducted remotely during the early months of COVID-19. I especially benefited from multiple sessions with John Seely Brown and with Lucy

Suchman, as well as an extensive session with Natalie Jeremijenko (we may have set a record for longest continuous intercontinental Zoom call).

Discussing the project with my colleagues and fellow writers was also essential. Fred Turner tops that list. (Authors who write about the history of computing are always thanking Fred.) The generosity with which he shares his greatness—with anyone who knocks on his office door—is a rare and beautiful thing. Fred sponsored my appointment as a visiting scholar at Stanford, and he gave me such energizing feedback at crucial intervals. I am grateful for conversations with many others, too, including but not limited to Leslie Berlin, Cameron Blevins, Stephen Derksen, Katherine Goodman, Jacob Greene, Sarah Hagelin, Amy Hasinoff, Pamela Laird, Elizabeth Losh, Joanna Luloff, Sean Morey, Gillian Silverman, Dale Stahl, Esther Sullivan, and Leila Takayama. Thank you to the community at Denver's Lighthouse Writers Workshop; my prose benefited from narrative nonfiction workshops led by amazing authors like Sarah M. Broom, Steve Knopper, and Helen Thorpe.

Funding was provided by a grant from the Office of Research Services at the University of Colorado Denver, and by travel funds from the College of Liberal Arts and Sciences and the English Department, where I am proud to work. Two undergraduate English majors, Lucas Duddles and Kira Morris, provided excellent research assistance—bright futures await them both.

Deirdre Mullane, literary agent extraordinaire, seems to know exactly what to tell me at every step in the process. I owe a lot to her guidance, encouragement, and close readings. Every writer should be so lucky. Joseph Calamia, science and technology editor at Chicago, was the editor I'd been hoping to work with from the start. He has proven to be the closest thing I've known to Maxwell Perkins. Again, every writer should be so lucky. The manuscript also benefited from astute suggestions offered by three anonymous peer reviewers and from the incredible copyediting done by Johanna Rosenbohm. Magda Wojcik provided wonderful proofreading. I thank Anjali Anand and Elizabeth Ellingboe at Chicago for guiding the book into production, as well as Carrie Olivia Adams for leading the publicity charge.

Most of all, thank you to Kim and Hutton, who sacrificed six walks in the woods to help me compile endnotes—and for so much else. Thank you for always being there for me through it all.

NOTES

Prologue

1 Wolkomir, "Computers Coming Out of the Woodwork," 82.
2 Sullivan, "'Calm Computing' Creator Dies at 46."
3 Bill Gates, memo to executive staff, August 28, 1991, box 1, folder 9, Mark D. Weiser Papers, M1069, Department of Special Collections, Stanford University Libraries, Stanford, California (hereafter referred to as Mark D. Weiser Papers).
4 Weiser, "The Computer for the 21st Century," 104.
5 Ibid., 94.
6 Jobs, "When We Invented the Personal Computer."
7 Mark Weiser, "Workshop on Ubiquitous Computing: What's Next," April 1993, box 75, folder 14. Mark D. Weiser Papers.
8 Weiser, "Open House."
9 Weiser, "The World Is Not a Desktop," 8.
10 Mark Weiser, abstract for keynote lecture on ubiquitous computing at University of British Columbia, November 1993, box 75, folder 15, Mark D. Weiser Papers.

Introduction

1 See Mandeep and Rajni, "A Systematic Study of Load Balancing Approaches," 9203.
2 "Smart Refrigerator Market."
3 Browning, "IoT Started with a Vending Machine."
4 Bogost, "The Internet of Things You Don't Really Need."
5 Badger, "Google's Founders Wanted to Shape a City."
6 Garfield, "Google's Parent Company Is Spending $50 Million."
7 Tierney, "Toronto's Smart City," 11.
8 See Corkery and Silver-Greenberg, "Miss a Payment?"

9 Sadowski, *Too Smart*, 123. See also Halpern, Mitchell, and Geoghegan, "The Smartness Mandate." Citing IBM's 2008 "Smarter Planet" ad campaign as well as Songdo, South Korea (a smart city built in partnership with Cisco), Halpern and colleagues distilled a set of principles that corporate smart city initiatives generally share. Proponents of the smartness mandate believe, Halpern observed, that intelligence emerges from studying big groups—the larger the population, the greater the yield. In turn, "a key premise of smartness is that while each member of a population is unique, it is also 'dumb'" (ibid., 117). The only individuals who might command any real insight, under this model, are those who can access a great deal of data about the rest of us.

10 McKenzie, "Sidewalk Labs Is Toronto's Best Hope."

11 Ibid.

12 Cavoukian, "De-identifying Data at the Source."

13 Ibid.

14 Zuboff, *Age of Surveillance Capitalism*, 78.

15 Doctoroff, "Google City," 4:40–5:31.

16 Zuboff, *Age of Surveillance Capitalism*, 10.

17 Zuboff, "Toronto Is Surveillance Capitalism's New Frontier."

18 Doctoroff, "Why We're No Longer Pursuing the Quayside Project."

19 Kirkwood and Scott, "Toronto Mayor Believes Labs Wanted 'Bargain Basement Price.'"

20 Zuboff, "Toronto Is Surveillance Capitalism's New Frontier."

21 Zuboff, *Age of Surveillance Capitalism*, 206.

22 ATLAS Institute, "Alan Kay Speaks [. . . Polymaths Unite!]," 27:35–27:40.

23 Johan de Kleer, author interview, April 3, 2020.

24 See Smith and Alexander, *Fumbling the Future*.

25 Brin, "Why Google Glass?," 2:42–5:00; CNBC, "Google's Eric Schmidt: 'The Internet Will Disappear,'" 0:12.

26 CNBC, "Google's Eric Schmidt: 'The Internet Will Disappear,'" 0:12–0:28.

27 Paradiso, "Our Extended Sensoria."

28 Case, *Calm Technology*, viii.

29 Mark Weiser, presentation slides for lecture "21st Century Challenges for Distributed Computing," June 1994, box 66, folder 11, Mark D. Weiser Papers.

30 Weiser, "The Future of Ubiquitous Computing on Campus," 42.

31 Mumford, *Technics and Civilization*, 324.

Chapter One

1 For examples during the 1970s, see Hiltzik, *Dealers of Lightning*, which gives account of Adele Goldberg's and Diana Merry's pivotal contributions to the development of the Smalltalk programming language, as well as Lynn Conway's pioneering innovations in VLSI–chip design methodology.

Subsequent interdisciplinary advances at PARC during the 1980s were led by the mathematician and computer scientist Frances Yao, the color modeling and graphics expert Maureen C. Stone, and the anthropologist Lucy Suchman.

2 John Seely Brown, author interview, July 16, 2020.

3 "PARC History."

4 Schmidt quoted in "Xerox PARC Has New Tech Leader."

5 Brown, "The Debriefing."

6 John Seely Brown, author interview, March 18, 2020.

7 Ibid.

8 Kay, "The Dynabook—Past, Present, and Future," 42:55–43:02.

9 John Maxwell, author interview, September 16, 2020.

10 John Seely Brown, author interview, July 16, 2020.

11 Ibid.

12 John Seely Brown, author interview, July 16, 2020.

13 Brown, foreword to *Making Work Visible*, xxi.

14 The Turing Award winner Edward Feigenbaum is generally pointed to as "the father of expert systems," primarily on account of the pioneering Dendral project he and his research team initiated in 1965 at Stanford University.

15 Johan de Kleer, author interview, April 3, 2020.

16 Ibid.

17 Brown, foreword to *Making Work Visible*, xxii.

18 Ibid.

19 Ibid., xxiii.

20 Ibid.

21 Ibid.

22 Hiltzik, "2 Brothers' High-Tech History."

23 Brown, foreword to *Making Work Visible*, xxiii.

24 Ibid.

25 Brown intended this letter to be included as a sidebar to his widely read article "Research That Reinvents the Corporation," but the editors at *Harvard Business Review* decided against publishing the letter.

26 Brown, "Letter to a Young Researcher."

27 Ibid.

28 Ibid.

29 Ibid.

30 Ibid.

31 Vicky Reich, author interview, February 25, 2020.

32 Ibid.

33 Ibid.

34 Randy Trigg, author interview, April 25, 2020.

35 Mark Weiser to Sun Microsystems, April 7, 1986, box 1, folder 4, Mark D. Weiser Papers.

36 Mark Weiser to United Airlines, November 20, 1986, box 1, folder 6, Mark D. Weiser Papers.
37 Bush, "As We May Think."
38 Engelbart, "Augmenting Human Intellect."
39 Turner, *From Counterculture to Cyberculture*, 107.
40 Victor, "A Few Words on Doug Engelbart."
41 Levy, *Insanely Great*, 42.
42 Waldrop, *The Dream Machine*, 367.
43 Thacker quoted in Bardini, *Bootstrapping*, 163.
44 Licklider quoted in Ohshima, "Some Reflections on Early History by Licklider," 16:18–17:10.
45 Ibid.
46 Kay, "The Dynabook—Past, Present, and Future," 36:45–36:50.
47 Ibid., 44:20–44:34.
48 Goldberg, *History of Personal Workstations*, viii.
49 Vicky Reich, author interview, February 25, 2020.
50 Corinne Reich-Weiser, author interview, July 8, 2021.
51 Ibid.
52 Vicky Reich, author interview, February 25, 2020.
53 Ibid.
54 Brown quoted in Curiel, "Mark Weiser."
55 John Seely Brown, author interview, July 16, 2020.
56 Ibid.
57 Brown quoted in Sullivan, "'Calm Computing' Creator Dies at 46."
58 John Seely Brown, author interview, July 16, 2020.

Chapter Two

1 Mark Weiser, "Where I'm At and How I Got Here," 1970, box 143, folder 22, Mark D. Weiser Papers.
2 Ibid.
3 Vicky Reich, author interview, February 25, 2020.
4 David Weiser, journal entries, 1963–1977, box 180, folders 8–19, Mark D. Weiser Papers.
5 Audra Weiser, diary entry, January 22, 1969, box 179, folder 20, Mark D. Weiser Papers.
6 Gardner, *Hexaflexagons and Other Mathematical Diversions*, 35.
7 Mark Weiser, "Where I'm At and How I Got Here," 1970, box 143, folder 22, Mark D. Weiser Papers.
8 Vicky Reich, author interview, February 25, 2020.
9 Ibid.
10 Audra Weiser, diary entry, November 21, 1968, box 179, folder 20, Mark D. Weiser Papers.
11 Ibid.

12 David Weiser, syllabus for Chemistry 286, box 180, folder 6, Mark D. Weiser Papers.
13 Polanyi, *The Tacit Dimension*, 4.
14 Ibid., 4–5.
15 Weiser, "The Computer for the 21st Century," 94.
16 Polanyi, *The Tacit Dimension*, xviii.
17 Ibid., 12.
18 Ibid., 9.
19 "New College Educational Contract," 1971, box 143, folder 5, Mark D. Weiser Papers.
20 Huntley, "Would You Let Your Kid Go to New College?"
21 Vicky Reich, author interview, February 25, 2020.
22 See Lackey, "What Are the Modern Classics?," 331.
23 Mark Weiser, "Relevance of the Cartesian Understanding," 1969, box 143, folder 1, Mark D. Weiser Papers.
24 Mark Weiser, "A Way of Looking at Experience," 1969, box 143, folder 1, Mark D. Weiser Papers.
25 Mark Weiser, notebook at New College, fall 1970, box 143, folder 22, Mark D. Weiser Papers.
26 Ibid.
27 Heidegger, *Being and Time*, 98.
28 Mark Weiser, "Hello," September 23, 1970, box 143, folder 22, Mark D. Weiser Papers.
29 Mark Weiser, "I Want to Love Someone," October 19, 1970, box 143, folder 22, Mark D. Weiser Papers.
30 Mark Weiser, "Presto-Chango," October 12, 1970, box 143, folder 22, Mark D. Weiser Papers.
31 Mark Weiser to Dr. Riley of New College, January 20, 1973, box 143, folder 22, Mark D. Weiser Papers.
32 Weiser, "The Computer for the 21st Century," 94.
33 Mark Weiser, "To David from Mark," audio recording, 55:09, September 15, 1974, box 133, folder 19, Mark D. Weiser Papers.
34 Ibid.
35 Ibid.
36 Ibid.
37 Ibid.
38 Ibid.
39 Ibid.
40 Kauffman and New, *Co-counseling*, 2.
41 Carr, "Attack Theory," 6.
42 Reevalution Counseling, "A Brief History of RC."
43 Carr, "Attack Theory," 6.
44 Mark Weiser, notes about reevaluation counseling, 1978, box 157, folder 14, Mark D. Weiser Papers.

45 Handout about the fundamentals of reevaluation counseling, 1975, box 156, folder 7, Mark D. Weiser Papers.
46 Mark Weiser, notebook for reevaluation counseling, 1975, box 156, folder 7, Mark D. Weiser Papers.
47 Handout about the fundamentals of reevaluation counseling, 1975, box 156, folder 7, Mark D. Weiser Papers.
48 Jackins, "The Postulates of Reevaluation Counseling."
49 Jackins, *The Human Side of Human Beings*, 5–7.
50 Ibid., 5.
51 Ibid., 9.
52 Ibid., 11.
53 Ibid., 13.
54 Ibid.
55 Ibid.
56 Handout for teachers of reevaluation counseling, 1975, box 157, folder 9, Mark D. Weiser Papers.
57 Jackins, *The Human Side of Human Beings*, 16.
58 Mark Weiser, notes from a reevaluation counseling workshop session, 1975, box 157, folder 9, Mark D. Weiser Papers.
59 Topol, "A History of MTS," 13.
60 Jerry Hiniker quoted in Topol, "A History of MTS," 17.
61 "Welcome to the University of Michigan Computing Center."

Chapter Three

1 Mark Weiser, "Visioning Report #1," June 1998, box 31, folder 7, Mark D. Weiser Papers.
2 Mark Weiser and John White, "Building a CSL Vision," June 1998, box 31, folder 7, Mark D. Weiser Papers.
3 Mark Weiser, goals chart for reevaluation counseling, 1975, box 156, folder 7, Mark D. Weiser Papers.
4 Mark Weiser, "Towards an Integrated Long-Range Vision for CSL," June 1998, box 31, folder 7, Mark D. Weiser Papers.
5 Ibid.
6 Ibid.
7 Ibid.
8 Jean Gastinel et al., "The CopierLess Office: A Play on Technology," June 1998, box 31, folder 8, Mark D. Weiser Papers.
9 For all the project proposals, see box 31, folders 7–8, Mark D. Weiser Papers.
10 Mark Weiser, "What I've Learned from Visioning So Far," summer 1998, box 31, folder 7, Mark D. Weiser Papers.
11 Weiser, "The Origins of Ubiquitous Computing Research," 693.

12 Stefik et al., "Beyond the Chalkboard," 32.
13 Ibid.
14 Ibid.
15 Ibid., 33.
16 Stefik, "The Colab Movie (1987)," 0:21–0:29.
17 Stefik et al., "Beyond the Chalkboard," 34.
18 Ibid.
19 See Negroponte, "5 Predictions, from 1984."
20 Brand, *The Media Lab*, 139.
21 Bolt and Negroponte, *Spatial Data-Management*, 12.
22 See From the Vault of MIT, "Soft Machine," 1:45–3:42.
23 Ibid.
24 An important (though lesser-known) predecessor who pursued similar efforts at MIT during the 1960s was Warren M. Brodey; see Brodey, "The Design of Intelligent Environments."
25 Negroponte, *Soft Architecture Machines*, 133.
26 Ibid., 144.
27 Mark Weiser to Harvey Jackins, 1975, box 156, folder 15, Mark D. Weiser Papers.
28 Vicky Reich, author interview, February 25, 2020.
29 See Marcuse, *One-Dimensional Man*.
30 Suchman, "Conversation with Lucy Suchman."
31 See Suchman, *Plans and Situated Actions*.
32 Mark Weiser, interview for PAIR program at PARC, June 1993, box 6, folder 2, Mark D. Weiser Papers.
33 John Seely Brown, author interview, March 18, 2020.
34 Stanley, *Mothers and Daughters of Invention*, 505.
35 Suchman, *Plans and Situated Actions*, 110.
36 Dan Russell, author interview, April 13, 2020.
37 John Seely Brown, author interview, March 18, 2020.
38 See Burke, "Analysis of Intelligibility in a Practical Activity."
39 Suchman, *Plans and Situated Actions*, 71.
40 Suchman and Jordan, "Interactional Troubles in Face-to-Face Survey Interviews," 233.
41 See Suchman, *Plans and Situated Actions*, 122–64.
42 Ibid., 168.
43 Suchman, "Agencies at the Interface."
44 Suchman, *Plans and Situated Actions*, 169.
45 Ibid., 185.
46 Weiser, "The Origins of Ubiquitous Computing Research," 693.
47 Ibid.
48 Lucy Suchman, author interview, January 21, 2020.
49 Smith and Alexander, *Fumbling the Future*, 15.

50 "Fumbling the Future," book review in the *Economist*.
51 Mark Weiser, "Creating the Invisible Interface," 1994, box 66, folder 1, Mark D. Weiser Papers.
52 Smith and Alexander, *Fumbling the Future*, 56.
53 Mark Weiser, "20 Years Out," November 1988, box 31, folder 11, Mark D. Weiser Papers.
54 Weiser quoted in Wolkomir, "Computers Coming Out of the Woodwork," 85.
55 Weiser quoted in Wasserman, "Here, There and Everywhere."
56 Saffo quoted in Wasserman, "Here, There and Everywhere."
57 Mel et al., "TABLET: Personal Computer in the Year 2000," 639.
58 Ibid., 642.
59 Mark Weiser, "The Future of Computing," November 1988, box 31, folder 8, Mark D. Weiser Papers.
60 Mark Weiser, "Beyond Workstations," February 1989, box 61, folder 9, Mark D. Weiser Papers.

Chapter Four

1 Kaplan quoted in Markoff, "The Big News in Tiny Computers."
2 Markoff, "The Big News in Tiny Computers."
3 Mark Weiser, "Ubiquitous Computing," May 15, 1989, box 61, folder 11, Mark D. Weiser Papers.
4 Ibid.
5 Bateson, *Steps to an Ecology of Mind*, 457.
6 Mark Weiser, "Ubiquitous Computing," May 15, 1989, box 61, folder 11, Mark D. Weiser Papers.
7 Ibid.
8 Ibid.
9 Ibid.
10 Ibid.
11 Ibid.
12 Ibid.
13 Heidegger, *Martin Heidegger: Basic Writings*, 321.
14 Ibid.
15 Weiser, "The Computer for the 21st Century," 104.
16 Mark Weiser, "Ubiquitous Computing," May 15, 1989, box 61, folder 11, Mark D. Weiser Papers.
17 Gassee quoted in "Computing in the Year 2000," 38.
18 Weiser quoted in Furger, "PARC Scientists Strive to Pursue Radical Ideas."
19 Feinstein, "Making Computers Invisible," 28.
20 Leong, "Innovation in the Industry's Labs," 84.
21 Metcalfe quoted in "Computing in the Year 2000," 39.

22 John Seely Brown, performance appraisal for Mark Weiser, 1989, box 55, folder 9, Mark D. Weiser Papers.
23 Ibid.
24 Weiser, "The Technologist's Responsibilities and Social Change," 17.
25 Ibid.
26 Mark Weiser, "Pre-history of Ubi," April 1993, box 75, folder 14, Mark D. Weiser Papers.
27 Ibid.
28 Suchman, *Plans and Situated Actions*, 189.
29 Ibid.
30 See Sundblad, "UTOPIA: Participatory Design from Scandinavia."
31 Mark Weiser, "The Future of Computing," November 1988, box 31, folder 8, Mark D. Weiser Papers.
32 Mark Weiser et al., "X Pad Plan," September 1990, box 97, folder 15, Mark D. Weiser Papers.
33 Weiser, "The Technologist's Responsibilities and Social Change."
34 Mark Weiser, "Ubiquitous Computing," May 15, 1989, box 61, folder 11, Mark D. Weiser Papers.
35 Mark Weiser et al., "X Pad Plan," September 1990, box 97, folder 15, Mark D. Weiser Papers.
36 Elrod, Bruce, et al., "Liveboard," 600.
37 Streitz, "An Interview with Ubicomp Pioneer Norbert Streitz," 62.
38 See Kinsley, "Ubiquitous Computing—Xerox PARC Circa 1991." 0:26– 0:37.
39 Ibid.
40 Want et al., "Overview of the PARCTAB Experiment," 16.
41 Ibid., 17.
42 Ibid., 19.
43 Rich Gold to John Seely Brown, October 8, 1990, box 36, folder 12, Mark D. Weiser Papers.
44 Vicky Reich, author interview, February 25, 2020.
45 Ibid.
46 Corinne Reich-Weiser, author interview, July 8, 2021.
47 Nicole Reich-Weiser, author interview, June 17, 2021.
48 See Lewis, *The New New Thing*.
49 Clark, "A TeleComputer," 23.
50 Berners-Lee, "World Wide Web."
51 Curtis, "Mudding," 140.
52 Curtis quoted in Reed, "1990: LambdaMOO."
53 Curtis quoted in Evans, "A Mansion Filled with Hidden Worlds."
54 Curtis, "Mudding," 121.
55 Weiser, "The Computer for the 21st Century," 104.
56 Mark Weiser, notebook entry, 1990, box 59, folder 5, Mark D. Weiser Papers.

57 Pam Roderick, author interview, April 23, 2020.
58 Mark Weiser, notebook entry, 1990, box 59, folder 5, Mark D. Weiser Papers.
59 Vicky Reich, author interview, February 25, 2020.
60 Mark Weiser, notebook entry, 1990, box 59, folder 5, Mark D. Weiser Papers.
61 Ibid.
62 Nicole Reich-Weiser, author interview, June 17, 2021.
63 Scott Elrod, author interview, May 26, 2020.

Chapter Five

1 Bill Gates, memo to executive staff, August 28, 1991, box 1, folder 9, Mark D. Weiser Papers.
2 John Seely Brown, author interview, March 18, 2020.
3 Ibid.
4 Assorted postcards to Weiser, reprint requests, fall 1991, box 1, folder 11, Mark D. Weiser Papers.
5 Takayama, "The Motivations of Ubiquitous Computing," 558.
6 Markoff, "Not a Personal Computer in Sight."
7 See Van, "Invisibility May Be Next for Technology"; and Potts, "Rethinking Computers, Companies."
8 Weiser, "The Computer for the 21st Century," 94.
9 Wells, *World Brain*, 56.
10 Vicky Reich, author interview, February 25, 2020.
11 Weiser, "The Computer for the 21st Century," 94.
12 Ibid.
13 Ibid.
14 Ibid.
15 Ibid.
16 Ibid.
17 Ibid., 94, 104.
18 Ibid., 98.
19 Ibid., 100.
20 Ibid.
21 Ibid., 104.
22 Ibid., 102.
23 Ibid.
24 Ibid.
25 Ibid.
26 Ibid.
27 Ibid.
28 Ibid.
29 Ibid., 104.

30 Ibid.

31 Vicky Reich, author interview, February 25, 2020.

32 Negroponte quoted in Brody, "Machine Dreams," 33.

33 "Interface Agents," *Frames: A Monthly Publication for Sponsors of the Media Lab*, November 1992, box 41, folder 3, Mark D. Weiser Papers.

34 Ibid.

35 Weiser quoted in ibid. See also Weiser, "Open House."

36 Tesler, "Networked Computing in the 1990s," 88.

37 Kay quoted in "Interface Agents," *Frames: A Monthly Publication for Sponsors of the Media Lab*, November 1992, box 41, folder 3, Mark D. Weiser Papers.

38 Kay, "Computers, Networks and Education," 146.

39 Maes quoted in "Interface Agents," *Frames: A Monthly Publication for Sponsors of the Media Lab*, November 1992, box 41, folder 3, Mark D. Weiser Papers.

40 Ibid.

41 Negroponte, "Products and Services for Computer Networks," 111.

42 See Zuboff, *Age of Surveillance Capitalism*, 63–97.

43 Weiser quoted in Sandberg, "Datorn Försvinner: Interview with Mark Weiser."

44 Weiser, "The Computer for the 21st Century," 104.

45 Ibid.

46 Weiser quoted in Van, "Invisibility May Be Next for Technology."

47 Weiser quoted in "Interface Agents," *Frames: A Monthly Publication for Sponsors of the Media Lab*, November 1992, box 41, folder 3, Mark D. Weiser Papers.

48 Mark Weiser, "Does Ubiquitous Computing Need Interface Agents?," presentation, October 1992, box 63, folder 12, Mark D. Weiser Papers.

49 Ibid.

50 Ibid.

51 Ibid.

52 Weiser, "The Computer for the 21st Century," 94.

53 Weiser quoted in "Interface Agents," *Frames: A Monthly Publication for Sponsors of the Media Lab*, November 1992, box 41, folder 3, Mark D. Weiser Papers.

54 Ibid.

55 Ibid.

56 Meyer Billmers quoted in "Interface Agents," *Frames: A Monthly Publication for Sponsors of the Media Lab*, November 1992, box 41, folder 3, Mark D. Weiser Papers.

57 Negroponte quoted in "Interface Agents," *Frames: A Monthly Publication for Sponsors of the Media Lab*, November 1992, box 41, folder 3, Mark D. Weiser Papers.

58 Ibid.

59 Weiser quoted in "Interface Agents," *Frames: A Monthly Publication for Sponsors of the Media Lab*, November 1992, box 41, folder 3, Mark D. Weiser Papers.

Chapter Six

1 Mark Weiser, "Workshop on Ubiquitous Computing: What's Next," April 1993, box 75, folder 14, Mark D. Weiser Papers.
2 "Where the Future Is Being Designed."
3 Mark Weiser, "Workshop on Ubiquitous Computing: What's Next," April 1993, box 75, folder 14, Mark D. Weiser Papers.
4 Ibid.
5 Ibid.
6 Ibid.
7 Berntsen, Elgsaas, and Hegna, "The Many Dimensions of Kristen Nygaard," 43.
8 Eric Saund, trip report from PARC workshop on ubiquitous computing, April 1993, box 52, folder 17, Mark D. Weiser Papers.
9 John Ellis, position paper for PARC workshop on ubiquitous computing, April 1993, box 52, folder 18, Mark D. Weiser Papers.
10 Brent Welch, position paper for PARC workshop on ubiquitous computing, April 1993, box 52, folder 18, Mark D. Weiser Papers.
11 Rich Gold, position paper for PARC workshop on ubiquitous computing, April 1993, box 52, folder 18, Mark D. Weiser Papers.
12 Ibid.
13 Gitti Jordan, trip report from PARC workshop on ubiquitous computing, April 1993, box 52, folder 17, Mark D. Weiser Papers.
14 Lucy Suchman, position paper for PARC workshop on ubiquitous computing, April 1993, box 52, folder 18, Mark D. Weiser Papers.
15 Gitti Jordan, position paper for PARC workshop on ubiquitous computing, April 1993, box 52, folder 18, Mark D. Weiser Papers.
16 Ibid.
17 Lucy Suchman, position paper for PARC workshop on ubiquitous computing, April 1993, box 52, folder 18, Mark D. Weiser Papers.
18 Gitti Jordan, position paper for PARC workshop on ubiquitous computing, April 1993, box 52, folder 18, Mark D. Weiser Papers.
19 Randy Trigg, position paper for PARC workshop on ubiquitous computing, April 1993, box 52, folder 18, Mark D. Weiser Papers.
20 Quoted in Eric Saund, trip report from PARC workshop on ubiquitous computing, April 1993, box 52, folder 17, Mark D. Weiser Papers.
21 Mark Weiser, notebook entry, 1991, box 58, folder 6, Mark D. Weiser Papers.
22 John A. Hughes, report on CSL, 1995, box 32, folder 1, Mark D. Weiser Papers.
23 Lucy Suchman, author interview, July 8, 2020.

24 Ibid.
25 Suchman, "Working Relations of Technology Production and Use."
26 Ibid., 27.
27 Ibid., 28.
28 Ibid.
29 Ibid., 29.
30 Ibid., 34.
31 Ibid.
32 Ibid.
33 Mark Weiser, "Towards a Project for the Lab Manager Lab," September 1992, box 2, folder 7, Mark D. Weiser Papers.
34 See Albrecht, "The Evolution of the Digital Mind."
35 Mark Weiser, notebook entry, 1992, box 58, folder 6, Mark D. Weiser Papers.
36 Mark Weiser, "Reply to Bob Anderson's Comments on Sandra Harding," October 1992, box 2, folder 6, Mark D. Weiser Papers.
37 Mark Weiser, interview for PAIR program at PARC, June 1993, box 6, folder 2, Mark D. Weiser Papers.
38 Mark Weiser, notebook entry, 1992, box 58, folder 6, Mark D. Weiser Papers.
39 Ibid.
40 Wolkomir, "Computers Coming Out of the Woodwork," 86.
41 Waldrop, "PARC Builds a World," 1523.
42 Elrod, Hall, et al., "Responsive Office Environments," 84.
43 Gold, "This Is Not a Pipe," 72.
44 Rich Gold, position paper for PARC workshop on ubiquitous computing, April 1993, box 52, folder 18, Mark D. Weiser Papers.
45 Ibid.
46 Ibid.
47 Weiser, "The World Is Not a Desktop," 8.
48 John Seely Brown, "Making the Future Possible: An Interview with John Seely Brown," n.d., box 32, folder 16, Mark D. Weiser Papers.
49 Ibid.
50 John Seely Brown, author interview, July 16, 2020.
51 Rheingold, "PARC Is Back!"
52 Clement, "Considering Privacy in Multi-media Communications."
53 Weiser, "The Technologist's Responsibilities and Social Change," 17.
54 Want quoted in Sloane, "Orwellian Dream Come True: A Badge That Pinpoints You."
55 Weiser quoted in Bozman and Booker, "Looking Forward to Office 2001," 28.
56 Lucy Suchman, author interview, July 8, 2020.
57 Theimer quoted in Wolkomir, "Computers Coming Out of the Woodwork," 89.
58 Lucy Suchman, author interview, July 8, 2020.
59 Ibid.

60 Weiser, "The World Is Not a Desktop," 7.
61 Wolkomir, "Computers Coming Out of the Woodwork," 82.
62 John Seely Brown, author interview, July 16, 2020.
63 Vicky Reich, author interview, February 25, 2020.
64 Mark Weiser, notebook entry, 1994, box 58, folder 5, Mark D. Weiser Papers.

Chapter Seven

1 Blomberg, Suchman, and Trigg, "Reflections on a Work-Oriented Design Project," 243.
2 Lucy Suchman, author interview, July 8, 2020.
3 Laird, "The Business of Consumer Culture History."
4 Dan Russell, author interview, April 13, 2020.
5 Eric Saund, email to author, February 23, 2020.
6 Zakon, "Hobbes' Internet Timeline 25."
7 Eric Saund, email to author, February 23, 2020.
8 Lewis, *The New New Thing*, 106.
9 Clark and Edwards, *Netscape Time*, 63.
10 McCullough, *How the Internet Happened*, 24.
11 Ibid.
12 Lewis, *The New New Thing*, 85.
13 Gates, "The Internet Tidal Wave."
14 Ibid.
15 Ibid.
16 Strauss, "Rolling Stones Live on Internet."
17 Ibid.
18 Leiby, "Mick Jagger's Overbyte."
19 Barlow, "A Declaration of the Independence of Cyberspace."
20 Weiser quoted in Kupfer, "Alone Together," 100.
21 Mark Weiser, Tacit Inc. presentation, 1995, box 51, folder 15, Mark D. Weiser Papers.
22 Mark Weiser, notebook entry, 1995, box 58, folder 7, Mark D. Weiser Papers.
23 Ibid.
24 Mark Weiser to Bob Gunderson and Tom Villeneuve, regarding Tacit Inc., October 1995, box 51, folder 13, Mark D. Weiser Papers.
25 Mark Weiser, notebook entry, 1995, box 58, folder 7, Mark D. Weiser Papers.
26 Ibid.
27 Mark Weiser, "Tacit Inc Business Plan," November 1995, box 16, folder 12, Mark D. Weiser Papers.
28 Ibid.

29 McCullough, *How the Internet Happened*, 85.
30 "Tacit Inc Focus Group," audio recording, 1995, box 135, folder 4, Mark D. Weiser Papers.
31 Ibid.
32 Ibid.
33 Ibid.
34 Mark Weiser, notebook entry, 1995, box 58, folder 7, Mark D. Weiser Papers.
35 "Tacit Inc Focus Group," audio recording, 1995, box 135, folder 4, Mark D. Weiser Papers.
36 VC firm rejection letters for Tacit Inc., 1996, box 14, folder 17, Mark D. Weiser Papers.
37 Mark Weiser, timeline with projections for Tacit Inc., 1995, box 51, folder 13, Mark D. Weiser Papers.
38 Mark Weiser, notebook entry, 1994, box 58, folder 5, Mark D. Weiser Papers.
39 David Weiser, journal entry, January 27, 1964, box 180, folder 8, Mark D. Weiser Papers.
40 Ibid.
41 See Mark Weiser, "Lessons from a Silent Startup," 1996, box 59, folder 11, Mark D. Weiser Papers.
42 Negroponte, *Being Digital*, 50.
43 See Florman, "He Has Seen the Future and It Works."
44 Negroponte, *Being Digital*, 4.
45 Ibid., 7, 101, 145.
46 Ibid., 6.
47 Ibid., 148.

Chapter Eight

1 Mark Weiser, "Space in CSL and PARC," May 1990, box 97, folder 1, Mark D. Weiser Papers.
2 Weiser and Brown, "The Coming Age of Calm Technology," 83.
3 Natalie Jeremijenko, author interview, November 5, 2020.
4 Ibid.
5 Ibid.
6 Ibid.
7 Ibid.
8 Ibid.
9 Ibid.
10 Ibid.
11 Ibid.
12 Ibid.

13 Weiser and Brown, “The Coming Age of Calm Technology,” 77.
14 Gold, *The Plenitude*, 206.
15 Weiser and Brown, “Designing Calm Technology.”
16 Xerox revenues declined both steadily and sharply as the 1990s progressed. Their share of the copier market had fallen more and more since 1975, when an antitrust suit brought by the US government forced Xerox to hand over its patents for free to its competitors, including Japanese companies whose costs were far lower. The situation worsened as the internet went mainstream. While the rise of personal computing throughout the ’80s had boasted profits associated with printing and photocopying, the spread of email and online messaging platforms accelerated by AOL, Netscape, and subsequent dot-coms hit Xerox especially hard starting in the mid-’90s. See Day, “What Xerox Should Copy.”
17 Brown, “Calm Tech, Then and Now.”
18 John Seely Brown, author interview, July 16, 2020.
19 Ibid.
20 Weiser quoted in Xerox, “Xerox Names Computing Pioneer as Chief Technologist.”
21 LaBarre, “An Idea Factory for Industry,” 43.
22 Negroponte quoted in Markoff, “Information Technology.”
23 Gershenfeld, *When Things Start to Think*, 37.
24 Gershenfeld quoted in LaBarre, “An Idea Factory for Industry,” 43.
25 Rae-Dupree, “Gadget Uses Human Body.”
26 Guterl, “Reinventing the PC.”
27 Rae-Dupree, “Gadget Uses Human Body.”
28 Pentland, “Smart Rooms,” 71.
29 Ibid.
30 Picard, *Affective Computing*, 174.
31 Gold, *The Plenitude*, 57.
32 Ibid., 71.
33 Markoff, “Information Technology.”
34 Gold, *The Plenitude*, 59.
35 Ibid., 58.
36 Weiser quoted in Rae-Dupree, “Gadget Uses Human Body.”
37 Hawley quoted in LaBarre, “An Idea Factory for Industry,” 43.
38 Weiser and Brown, “The Coming Age of Calm Technology,” 75.
39 Ibid., 77.
40 Ibid., 79.
41 Ibid.
42 Ibid.
43 Ibid.
44 Ibid.
45 Ibid., 80.

46 Ibid., 79.
47 Weiser and Brown, "Center and Periphery," 317.
48 Weiser and Brown, "The Coming Age of Calm Technology," 80.
49 Ibid.
50 Ibid., 81.

Chapter Nine

1 Agre quoted in Masís, "Making AI Philosophical Again," 58.
2 Agre, *Computation and Human Experience*, 13.
3 Agre and Horswill, "Lifeworld Analysis," 113. See also Newell and Simon, "GPS: A Program That Simulates Human Thought."
4 Agre, "The Soul Gained and Lost."
5 Mitch Wagner, email to Mark Weiser, November 12, 1996, box 40, folder 8, Mark D. Weiser Papers.
6 Mark Weiser, email to Mitch Wagner, November 13, 1996, box 40, folder 8, Mark D. Weiser Papers.
7 Wagner, "HAL Is Born," 72.
8 Usher, "The Third Wave," 113.
9 Gershenfeld, *When Things Start to Think*, 204.
10 Weiser, "Open House."
11 Lewis, *The New New Thing*, 30.
12 Ibid., 298.
13 Ibid., 173.
14 Ibid., 134.
15 Ibid., 178.
16 See ibid., 295–301.
17 Weiser quoted in Pitta, "The Soul of the New Machine."
18 Weiser quoted in Gold, *The Plenitude*, 208.
19 Weiser, "Open House."
20 Ibid.
21 Ishii, "In January 1997, I received an email from Weiser."
22 Ibid.
23 Mark Weiser, email to Ishii, January 27, 1997, box 106, folder 15, Mark D. Weiser Papers.
24 Ishii and Ullmer, "Tangible Bits," 2.
25 Ibid.
26 Ibid.
27 Ibid.
28 Ibid., 5.
29 Ibid., 6.
30 Ibid., 2.
31 "About Ars Electronica."

32 Mark Weiser, "Periphery and the Flesh Factor," June 1997, box 96, folder 4, Mark D. Weiser Papers.
33 Merleau-Ponty, *Phenomenology of Perception*, 106.
34 Mark Weiser, "Periphery and the Flesh Factor," June 1997, box 96, folder 4, Mark D. Weiser Papers.
35 Merleau-Ponty, *Phenomenology of Perception*, 474.
36 Mark Weiser, "Periphery and the Flesh Factor," June 1997, box 96, folder 4, Mark D. Weiser Papers.
37 Ibid.
38 Mark Weiser to Hiroshi Ishii, January 27, 1997, box 106, folder 15, Mark D. Weiser Papers.
39 Mark Weiser, "Periphery and the Flesh Factor," June 1997, box 96, folder 4, Mark D. Weiser Papers.
40 Ibid.
41 Healey quoted in Wright, "'Wearables' Combine Fashion, High-Tech."
42 Rhodes, "Brief History of Wearable Computing."
43 The thin strip of hardware that Starner and his collaborator Doug Platt used for the Lizzy was a modification of the Private Eye head-mounted display, which was first produced by the Reflection Technology company in 1989. In 2010, about twelve years after the Wearables gala at MIT, Google would recruit Starner to develop a new and improved version of the Lizzy, which Google called Project Glass.
44 Pentland, "Smart Rooms," 73.
45 Jebara et al., "Stochasticks."
46 Wright, "'Wearables' Combine Fashion, High-Tech."
47 Ibid.
48 Hillinger, "Life on the 'Loneliest Road.'"
49 Mark Weiser to his mother and his sister Mona, March 14, 1998, box 5, folder 9, Mark D. Weiser Papers.
50 Ibid.
51 Ibid.
52 Ibid.
53 Vicky Reich, author interview, February 25, 2020.
54 Weiser, "The Future of Ubiquitous Computing on Campus," 42.
55 Ibid.
56 Ibid.
57 Ibid.
58 Ibid.
59 Russell and Weiser, "The Future of Integrated Design," 276.
60 Ibid.
61 Ibid.
62 Mark Weiser to his mother and his sister Mona, March 14, 1998, box 5, folder 9, Mark D. Weiser Papers.

Chapter Ten

1 Mark Weiser, "The Distinctiveness of Being Human," lecture, recorded at the University of California, Berkeley, June 1998, box 133, folder 15, Mark D. Weiser Papers.
2 Ibid.
3 Weiser, "The Spirit of the Engineering Quest," 357.
4 Weiser, "Mark and Nicole's Greek Vacation."
5 Weiser, "The Spirit of the Engineering Quest," 355.
6 Ibid., 360.
7 Ibid., 355.
8 Mark Weiser, "To David from Mark," audio recording, 55:09, September 15, 1974, box 133, folder 19, Mark D. Weiser Papers.
9 Ibid.
10 Weiser, "The Spirit of the Engineering Quest," 356.
11 Ibid., 359.
12 Ibid.
13 Ibid., 360.
14 Ibid., 358.
15 Ibid., 355.
16 Ibid., 361.
17 Ibid., 359.
18 Ibid.
19 Ibid., 360.
20 Ibid., 359.
21 Ibid., 361.
22 Dan Russell, author interview, April 13, 2020.
23 Ibid.
24 Ibid.
25 Ibid.
26 Ibid.
27 Vicky Reich, author interview, February 25, 2020.
28 Weiser quoted in Sullivan, "'Calm Computing' Creator Dies at 46."
29 Vicky Reich, author interview, May 28, 2021.
30 Mark Weiser, "The Distinctiveness of Being Human," recorded lecture at the University of California, Berkeley, June 1998, box 133, folder 15, Mark D. Weiser Papers.
31 Weiser quoted in Sullivan, "'Calm Computing' Creator Dies at 46."
32 Sullivan, "'Calm Computing' Creator Dies at 46."
33 Vicky Reich, author interview, May 28, 2021.
34 Ibid.
35 Ibid.
36 Vicky Reich, author interview, May 28, 2021.

37 Nicole Reich-Weiser, author interview, June 17, 2021.
38 Corinne Reich-Weiser, author interview, July 8, 2021.
39 Vicky Reich, author interview, May 28, 2021.
40 Gold quoted in Bahl, "A Reflection on Mark Weiser," 1.

Epilogue

1 Gold, *The Plenitude*, 246.
2 Eric Saund, email message to author, February 23, 2020.
3 Lucy Suchman, author interview, July 8, 2020.
4 Suchman, "Anthropological Relocations and the Limits of Design," 12.
5 Lucy Suchman, author interview, July 8, 2020.
6 Ibid.
7 Ibid.
8 Craig Mudge, author interview, October 26, 2020.
9 Johan de Kleer, author interview, April 3, 2020.
10 Stefik, "Mark Stefik on Invention and Innovation," 1.
11 "IEEE Pervasive Computing: Homepage."
12 Satyanarayanan, "A Catalyst for Mobile and Ubiquitous Computing," 2.
13 Bell and Dourish, "Yesterday's Tomorrows," 133.
14 See Dourish, *Where the Action Is*.
15 Protecstar, "Steve Jobs MacWorld Keynote in 2007," 50:15–50:20.
16 Ibid.
17 Dourish, *Where the Action Is*, 2.
18 Risley, "Apple's iOS 9 Brings Improvements."
19 Zuboff, *Age of Surveillance Capitalism*, 417.
20 See Turow, *The Voice Catchers*.
21 Khatchadourian, "We Know How You Feel."
22 "About Affectiva."
23 Silver, "Patents Reveal How Facebook Wants to Capture Your Emotions."
24 Pentland, "With Big Data Comes Big Responsibility."
25 Zuboff, *Age of Surveillance Capitalism*, 398.
26 Center for Humane Technology. "Resources for Technologists."
27 Weiser, "Open House."
28 Ibid.

BIBLIOGRAPHY

"About Affectiva." Affectiva. February 22, 2022. https://www.affectiva.com/about-affectiva/.

"About Ars Electronica." Ars Electronica. September 20, 2020. https://ars.electronica.art/about/en/.

Agre, Philip. *Computation and Human Experience*. New York: Cambridge University Press, 1997.

———. "The Soul Gained and Lost: Artificial Intelligence as a Philosophical Project." Unpublished manuscript, 1995. https://pages.gseis.ucla.edu/faculty/agre/shr.html.

Agre, Philip, and Ian Horswill. "Lifeworld Analysis." *Journal of Artificial Intelligence Research* 6 (1997): 111–45.

Albrecht, James. "The Evolution of the Digital Mind." Unpublished manuscript, circa 1993. Uploaded to *Medium*, November 26, 2018. https://medium.com/@newalbrecht/the-evolution-of-the-digital-mind-feddc2c54719.

ATLAS Institute, University of Colorado Boulder. "Alan Kay Speaks [. . . Polymaths Unite!]." November 11, 2019. YouTube video, 1:42:46. https://www.youtube.com/watch?v=nOrdzDaPYV4.

Badger, Emily. "Google's Founders Wanted to Shape a City. Toronto Is Their Chance." *New York Times*, October 20, 2017.

Bahl, Victor. "A Reflection on Mark Weiser." *Mobile Computing and Communications Review* 3, no. 3 (1999): 1.

Bardini, Thierry. *Bootstrapping: Douglas Engelbart, Coevolution, and the Origins of Personal Computing*. Stanford, CA: Stanford University Press, 2000.

Barlow, John Perry. "A Declaration of the Independence of Cyberspace." Eff .org, February 8, 1996. https://www.eff.org/cyberspace-independence.

Bateson, Gregory. *Steps to an Ecology of Mind: Collected Essays in Anthropology, Psychiatry, Evolution, and Epistemology*. San Francisco: Chandler, 1972.

Bell, Genevieve, and Paul Dourish. "Yesterday's Tomorrows: Notes on Ubiquitous Computing's Dominant Vision." *Personal and Ubiquitous Computing* 11, no. 2 (February 2007): 133–43.

Berners-Lee, Tim. "World Wide Web." CERN. n.d. http://info.cern.ch /hypertext/WWW/TheProject.html.

Berntsen, Drude, Knut Elgsaas, and Havard Hegna. "The Many Dimensions of Kristen Nygaard: Creator of Object-Oriented Programming and the Scandinavian School of System Development." In *History of Computing: Learning from the Past*, edited by Arthur Tatnall, 38–49. Berlin: Springer, 2010.

Blomberg, Jeanette, Lucy Suchman, and Randall H. Trigg. "Reflections on a Work-Oriented Design Project." *Human-Computer Interaction* 11 (1996): 237–65.

Bogost, Ian. "The Internet of Things You Don't Really Need." *Atlantic*, June 23, 2015. https://www.theatlantic.com/technology/archive/2015 /06/the-internet-of-things-you-dont-really-need/396485/.

Bolt, Richard A., and Nicholas Negroponte. *Spatial Data-Management*. Architecture Machine Group. Cambridge, MA: MIT Press, 1979.

Bozman, Jean S., and Ellis Booker. "Looking Forward to Office 2001." *Computerworld*, January 11, 1993, 28.

Brand, Stewart. *The Media Lab: Inventing the Future at MIT*. New York: Viking, 1987.

Brin, Sergey. "Why Google Glass?" Filmed February 27, 2013. TED video, 7:02. https://www.ted.com/talks/sergey_brin_why_google_glass.

Brodey, Warren M. "The Design of Intelligent Environments: Soft Architecture." *Landscape* 17, no. 1 (Autumn 1967): 8–12.

Brody, Herb. "Machine Dreams: An Interview with Nicholas Negroponte." *Technology Review* 95, no. 1 (January 1, 1992): 33.

Brown, John Seely. "Calm Tech, Then and Now." Interview. *Re:form*.

August 11, 2014. https://medium.com/re-form/calm-tech-then-and-now-deddb05697cf.
———. "The Debriefing: John Seely Brown." Interview by *Wired* staff. *Wired*, August 1, 2000. https://www.wired.com/2000/08/brown/.
———. Foreword to *Making Work Visible: Ethnographically Grounded Case Studies of Work Practice*, edited by Margaret H. Szymanski and Jack Whalen, xxi–xxvi. New York: Cambridge University Press, 2011.
———. "Letter to a Young Researcher." Unpublished manuscript, 1991, typescript.
———. "Research That Reinvents the Corporation." *Harvard Business Review* 69, 1 (January–February 1991): 102–11.
Browning, Daniel. "IoT Started with a Vending Machine." *Machine Design*, July 25, 2018. https://www.machinedesign.com/automation-iiot/article/21836968/iot-started-with-a-vending-machine.
Burke, Julie. "An Analysis of Intelligibility in a Practical Activity." PhD diss., University of Illinois Urbana–Champaign, 1982.
Bush, Vannevar. "As We May Think." *Atlantic Monthly*, July 1945. https://www.theatlantic.com/magazine/archive/1945/07/as-we-may-think/303881/.
Carr, Steve. "Attack Theory: Reevaluating RC." *Polemicist* 3, no. 5 (April 1992): 3–7.
Case, Amber. *Calm Technology: Principles and Patterns for Non-Intrusive Design*. Boston: O'Reilly, 2015.
Cavoukian, Ann. "De-identifying Data at the Source Is the Only Way Sidewalk Can Work." *Toronto Life*, September 4, 2019. https://torontolife.com/city/de-identifying-data-at-the-source-is-the-only-way-sidewalk-can-work/.
Center for Humane Technology. "Resources for Technologists." February 25, 2022. https://www.humanetech.com/technologists.
Clark, Jim. "A TeleComputer." *ACM SIGGRAPH Computer Graphics* 26, no. 2 (July 1992): 19–23.
Clark, Jim, and Owen Edwards. *Netscape Time: The Making of the Billion-Dollar Start-Up That Took on Microsoft*. New York: St. Martin's Press, 1999.

Clement, Andrew. "Considering Privacy in the Development of Multimedia Communications." *Computer Supported Cooperative Work* 2 (March 1993): 67–88.

CNBC. "Google's Eric Schmidt: 'The Internet Will Disappear.'" January 23, 2015. YouTube video, 2:29. https://www.youtube.com/watch?v=Tf49T45GNdo.

"Computing in the Year 2000." *InfoWorld*, January 2, 1989.

Corkery, Michael, and Jessica Silver-Greenberg. "Miss a Payment? Good Luck Moving That Car." *New York Times*, September 24, 2014.

Curiel, Jonathan. "Mark Weiser." *San Francisco Gate*, May 1, 1999. https://www.sfgate.com/news/article/Mark-Weiser-2933466.php.

Curtis, Pavel. "Mudding: Social Phenomena in Text-Based Virtual Realities." In *Culture of the Internet*, edited by Sara Kiesler, 121–42. New York: Psychology Press, 1997.

Day, George. "What Xerox Should Copy, and Not Copy, from Its Past." *Knowledge at Wharton*, October 25, 2000. https://knowledge.wharton.upenn.edu/article/what-xerox-should-copy-and-not-copy-from-its-past/.

Doctoroff, Daniel. "Google City: How the Tech Juggernaut Is Reimagining Cities—Faster Than You Realize." October 17, 2016. YouTube video, 27:40. https://www.youtube.com/watch?v=JXN9QHHD8eA.

———. "Why We're No Longer Pursuing the Quayside Project—and What's Next for Sidewalk Labs." *Medium*, May 7, 2020. https://medium.com/sidewalk-talk/why-were-no-longer-pursuing-the-quayside-project-and-what-s-next-for-sidewalk-labs-9a61de3fee3a.

Dourish, Paul. *Where the Action Is: The Foundations of Embodied Interaction*. Cambridge, MA: MIT Press, 2001.

Elrod, Scott, Richard Bruce, Rich Gold, David Goldberg, Frank Halasz, Willian Janssen, David Lee, Kim McCall, Elin Pedersen, Ken Pier, John Tang, and Brent Welch. "Liveboard: A Large Interactive Display Supporting Group Meetings, Presentations and Remote Collaboration." In *CHI '92: Proceedings of the Conference on Human Factors in Computing Systems*, 599–607. New York: ACM Press, 1992.

Elrod, Scott, Gene Hall, Rick Costanza, Michael Dixon, and Jim Des Rivieres. "Responsive Office Environments." *Communications of the ACM* 36, no. 7 (July 1993): 84–85.

Engelbart, Douglas C. "Augmenting Human Intellect: A Conceptual Framework." SRI Summary Report AFOSR-3223, October 1962. https://www.dougengelbart.org/content/view/138/.

Evans, Claire L. "A Mansion Filled with Hidden Worlds: When the Internet Was Young." *Undark*, July 20, 2018. https://undark.org/2018/07/20/wilo-evans-broad-band/.

Feinstein, Debra. "Making Computers Invisible." *Benchmark* 6, no. 3 (Fall 1989): 26–31.

Florman, Samuel C. "He Has Seen the Future and It Works: Review of *Being Digital* by Nicholas Negroponte." *New York Times Book Review*, February 5, 1995.

From the Vault of MIT. "Soft Machine—Architecture Machine Group (1984)." August 6, 2013. YouTube video, 8:47, https://www.youtube.com/watch?v=YOIWlgnM1Vs.

"Fumbling the Future." Review of *Fumbling the Future: How Xerox Invented, Then Ignored, the First Personal Computer*, by Douglas K. Smith and Robert C. Alexander, *Economist* 309, no. 7577 (November 19, 1988): 102.

Furger, Roberta. "PARC Scientists Strive to Pursue Radical Ideas." *InfoWorld*, July 3, 1989.

Gardner, Martin. *Hexaflexagons and Other Mathematical Diversions: The First Scientific American Book of Puzzles and Games*. Chicago: University of Chicago Press, 1988.

Garfield, Leanna. "Google's Parent Company Is Spending $50 Million to Build a High-Tech Neighborhood in Toronto." *Business Insider*, October 17, 2017. https://www.businessinsider.com/alphabet-sidewalk-labs-development-toronto-2017-10.

Gates, Bill. "The Internet Tidal Wave." Memo. May 26, 1995. Published on Wired.com. May 26, 2010. https://www.wired.com/2010/05/0526bill-gates-internet-memo/.

Gershenfeld, Neil A. *When Things Start to Think*. New York: Henry Holt, 1999.

Gold, Rich. *The Plenitude: Creativity, Innovation, and Making Stuff*. Cambridge, MA: MIT Press, 2007.

———. "This Is Not a Pipe." *Communications of the ACM* 36, no. 7 (July 1993): 72.

Goldberg, Adele. *A History of Personal Workstations*. New York: Association for Computing Machinery, 1988.

Guterl, Fred. "Reinventing the PC." *Discover Magazine*, August 31, 1995. https://www.discovermagazine.com/technology/reinventing-the-pc.

Halpern, Orit, Robert Mitchell, and Bernard Dionysius Geoghegan. "The Smartness Mandate: Notes Toward a Critique." *Grey Room* 68 (Summer 2017): 106–29.

Heidegger, Martin. *Being and Time*. Translated by John Macquarrie and Edward Robinson. New York: Harper Perennial, 1962.

———. *Martin Heidegger: Basic Writings*. Edited by David Farrell Krell. San Francisco: Harper, 1993.

Hillinger, Charles. "Life on the 'Loneliest Road.'" *Los Angeles Times*, August 25, 1986.

Hiltzik, Michael. *Dealers of Lightning: Xerox PARC and the Dawn of the Computer Age*. New York: HarperCollins, 1999.

———. "2 Brothers' High-Tech History in California." *Los Angeles Times*, February 19, 2004. https://www.latimes.com/archives/la-xpm-2004-feb-19-fi-golden19-story.html.

Huntley, Helen. "Would You Let Your Kid Go to New College?" *St. Petersburg Times*, November 14, 1971.

"IEEE Pervasive Computing: Homepage." IEEE.org, December 3, 2021. https://ieeexplore-ieee-org.aurarialibrary.idm.oclc.org/xpl/RecentIssue.jsp?punumber=7756.

Ishii, Hiroshi. 2018. "In January 1997, I received an email from the late Dr. Mark Weiser." Facebook, November 5, 2018. https://www.facebook.com/ishii.mit/posts/pfbid02Hnd8t8PLrLZhpPfakyR43m9Pn1KpkBVghXo2arSNM7u8iGqDgz4VwjKafgTRZFGhl?__cft__[0]=AZXASiRdDZEEolGwHgtiSoVlVFKoEZqznpCdpFIj2A8GFh_q8osloBj55NoMnQ5SxkpQW_LFVqrbfzo88rAY6Zjdo7cefuBqlRBI-JEQX_gdOf70TCsMYL5kIucVjMIhEzw&__tn__=%2CO%2CP-R.

Ishii, Hiroshi, and Brygg Ullmer. "Tangible Bits: Towards Seamless Interfaces between People, Bits and Atoms." In *CHI '97: Proceedings of the Conference on Human Factors in Computing Systems*. New York: ACM Press, 1997.

Jackins, Harvey. "The Postulates of Reevaluation Counseling." Reevaluation Counseling. Pamphlet, 1964. https://www.rc.org/publication/book/hs/postulates.

———. *The Human Side of Human Beings: The Theory of Re-evaluation Counseling*. Seattle: Rational Island Publishers, 1965.

Jebara, Tony, Cyrus Eyster, Josh Weaver, Thad Starner, and Alex Pentland. "Stochasticks: Augmenting the Billiards Experience with Probabilistic Vision and Wearable Computers." *Digest of Papers: First International Symposium on Wearable Computers* (1997): 138–45.

Jobs, Steve. "When We Invented the Personal Computer, We Created a New Kind of Bicycle." *Wall Street Journal*, August 13, 1980.

Kauffman, Katie, and Caroline New. *Co-counseling: The Theory and Practice of Reevaluation Counseling*. New York: Routledge, 2004.

Kay, Alan. "Computers, Networks and Education." *Scientific American* 265, no. 3 (September 1991): 138–48.

———. "The Dynabook—Past, Present, and Future." Computer History Museum. January 11, 2016. YouTube video, 1:51:53. https://www.youtube.com/watch?v=GMDphyKrAE8.

Khatchadourian, Raffi. "We Know How You Feel." *New Yorker*, January 12, 2015. https://www.newyorker.com/magazine/2015/01/19/know-feel.

Kinsley, Sam. "Ubiquitous Computing—Xerox PARC Circa 1991." May 28, 2009. YouTube video, 9:50. https://www.youtube.com/watch?v=b1w9_cob_zw.

Kirkwood, Isabelle, and Josh Scott. "Toronto Mayor Believes Sidewalk Labs Wanted 'Bargain Basement Price' for Quayside." Betakit.com, April 21, 2021. https://betakit.com/toronto-mayor-believes-sidewalk-labs-wanted-bargain-basement-price-for-quayside-hootsuite-ceo-pro-social-regulation-collision-2021/.

Kupfer, Andrew. "Alone Together: Will Being Wired Set Us Free?" *Fortune*, March 20, 1995, 94–104.

LaBarre, Polly. "An Idea Factory for Industry." *Industry Week*, February 19, 1996, 43.

Lackey, Douglas. "What Are the Modern Classics? The Baruch Poll of Great Philosophy in the Twentieth Century." *Philosophical Forum XXX*, no. 4 (December 1999): 329–46.

Laird, Pamela Walker. "The Business of Consumer Culture History: Systems, Interactions, and Modernization." In *Decoding Modern Consumer Societies*, edited by Hartmut Berghoff and Uwe Spiekermann, 89–110. New York: Palgrave Macmillan, 2012.

Leiby, Richard. "Mick Jagger's Overbyte." *Washington Post*, November 19, 1994.

Leong, Kathy Chin. "Innovation in the Industry's Labs and Think Tanks." *PC World* (September 1989): 84–87.

Levy, Steven. *Insanely Great: The Life and Times of Macintosh, the Computer That Changed Everything*. New York: Viking, 1994.

Lewis, Michael. *The New New Thing: A Silicon Valley Story*. New York: W. W. Norton, 1999.

Licklider, J. C. R. "Man-Computer Symbiosis." *IRE Transactions on Human Factors in Electronics* HFE-1, no. 1 (March 1960): 4–11.

Mandeep, Kaur, and Aron Rajni. "A Systematic Study of Load Balancing Approaches in the Fog Computing Environment." *Journal of Supercomputing* 77, no. 8 (2021): 9202–47.

Marcuse, Herbert. *One-Dimensional Man: Studies in the Ideology of Advanced Industrial Society*. Boston: Beacon, 1964.

Markoff, John. "The Big News in Tiny Computers." *New York Times*, May 14, 1989.

———. "Information Technology: And Now, Computerized Sensibility." *New York Times*, May 15, 1995.

———. "Not a Personal Computer in Sight." *New York Times*, October 6, 1991.

Masís, Jethro. "Making AI Philosophical Again: On Philip E. Agre's Legacy." *Continent* 4, no. 1 (2014): 58–70.

McCullough, Brian. *How the Internet Happened: From Netscape to the iPhone*. New York: Liveright Publishing Corporation, 2018.

McKenzie, Kwame. "Sidewalk Labs Is Toronto's Best Hope for Sustainability." *Toronto Life*, September 4, 2019. https://torontolife.com/city/sidewalk-labs-is-torontos-best-hope-for-sustainability/.

Mel, Bartlett W., Stephen M. Omohundro, Arch D. Robison, Steven S. Skiena, Kurt H. Thearling, and Luke T. Young. "TABLET: Personal Computer in the Year 2000." *Communications of the ACM* 31, no. 6 (June 1988): 638–48.

Merleau-Ponty, Maurice. *Phenomenology of Perception*. New York: Routledge, 1962.

Mumford, Lewis. *Technics and Civilization*. New York: Harcourt, 1934.

Negroponte, Nicholas. *Being Digital*. New York: Knopf, 1995.

———. "5 Predictions, from 1984." Filmed February 1984. TED video, 25:10. https://www.ted.com/talks/nicholas_negroponte_5_predictions_from_1984.

———. "Products and Services for Computer Networks." *Scientific American* 265, no. 3 (September 1991): 106–13.

———. *Soft Architecture Machines*. Cambridge, MA: MIT Press, 1975.

Newell, Allen, and H. A. Simon. "GPS: A Program That Simulates Human Thought." In *Computers and Thought*, edited by Edward Feigenbaum and Julian Feldman, 279–96. New York: McGraw-Hill, 1963.

Ohshima, Yoshiki. "Some Reflections on Early History by J.C.R. Licklider (VPRI 0093)." Filmed January 1986 in Palo Alto, California. YouTube video, 55:26. https://www.youtube.com/watch?v=SN--t9jXQco.

Paradiso, Joseph A. "Our Extended Sensoria: How Humans Will Connect with the Internet of Things." *MIT Technology Review*, August 1, 2017. https://www.technologyreview.com/2017/08/01/68061/our-extended-sensoria-how-humans-will-connect-with-the-internet-of-things/.

"PARC History." Parc.com, June 14, 2022. https://www.parc.com/about-parc/parc-history/.

Pentland, Alex P. "Smart Rooms." *Scientific American* 274, no. 4 (April 1996): 68–76.

———. "With Big Data Comes Big Responsibility: An Interview with Alex Pentland." *Harvard Business Review*, November 1, 2014. https://hbr.org/2014/11/with-big-data-comes-big-responsibility.

Picard, Rosalind. *Affective Computing*. Cambridge, MA: MIT Press, 1997.

Pitta, Julie. "The Soul of the New Machine." *Los Angeles Times*, November 18, 1996.

Polyani, Michael. *The Tacit Dimension*. Rev. ed. Chicago: University of Chicago Press, 2009.

Potts, Mark. "Rethinking Computers, Companies: Xerox Lab's Researchers See 'Ubiquitous Computing' Reshaping the Corporation." *Washington Post*, November 4, 1992.

Protecstar. "Steve Jobs MacWorld Keynote in 2007." May 16, 2013. YouTube video, 1:19:10. https://www.youtube.com/watch?v=VQKMoT-6XSg.

Rae-Dupree, Janet. "Gadget Uses Human Body to Exchange Electronic Business Cards." *San Jose Mercury News*, October 21, 1996.

Reed, Aaron A. "1990: LambdaMOO." 50 *Years of Text Games*, May 20, 2021. https://if50.substack.com/p/1990-lambdamoo.

Reevaluation Counseling. "A Brief History of RC." March 1, 2022. https://www.rc.org/publication/theory/origin.

Rheingold, Howard. "PARC Is Back!" *Wired*, February 1, 1994. https://www.wired.com/1994/02/parc/.

Rhodes, Bradley. "A Brief History of Wearable Computing." May 15, 2021. https://www.media.mit.edu/wearables/lizzy/timeline.html.

Risley, James. "Apple's iOS 9 Brings Improvements to Siri and New Context-Aware Features." *GeekWire*, June 8, 2015. https://www.geekwire.com/2015/ios-gets-smarter-with-improvements-to-siri-and-context-aware-features/.

Russell, Daniel M., and Mark Weiser. "The Future of Integrated Design of Ubiquitous Computing in Combined Real & Virtual Worlds." In *CHI '98: Proceedings of the Conference on Human Factors in Computing Systems*, 275–76. New York: ACM Press, 1998.

Sadowski, Jathan. *Too Smart: How Digital Capitalism Is Extracting Data, Controlling Our Lives, and Taking Over the World*. Cambridge, MA: MIT Press, 2020.

Satyanarayanan, Mahadev. "A Catalyst for Mobile and Ubiquitous Computing." *IEEE Pervasive Computing* 1, no. 1. (January–March 2002): 2–5.

Silver, Curtis. "Patents Reveal How Facebook Wants to Capture Your Emotions, Facial Expressions and Mood." Forbes.com, June 8, 2017. https://www.forbes.com/sites/curtissilver/2017/06/08/how-facebook-wants-to-capture-your-emotions-facial-expressions-and-mood/?sh=74d211a86014.

Sloane, Leonard. "Orwellian Dream Come True: A Badge That Pinpoints You." *New York Times*, September 12, 1992.

"Smart Refrigerator Market." Global Market Insights. https://www.gminsights.com/industry-analysis/smart-refrigerator-market.

Smith, Douglas K., and Robert C. Alexander. *Fumbling the Future: How Xerox Invented, Then Ignored, the First Personal Computer*. New York: W. Morrow, 1988.

Stanley, Autumn. *Mothers and Daughters of Invention: Notes for a Revised History of Technology*. New Brunswick, NJ: Rutgers University Press, 1995.

Stefik, Mark. "The Colab Movie (1987)." April 9, 2011. YouTube video, 3:23. https://www.youtube.com/watch?v=iPZTOosKjAE.

———. "Mark Stefik on Invention and Innovation." Interview. *Ubiquity*, November 2004, 1.

Stefik, Mark, Gregg Foster, Daniel G. Bobrow, Kenneth Kahn, Stan Lanning, and Lucy Suchman. "Beyond the Chalkboard: Computer Support for Collaboration and Problem Solving in Meetings." *Communications of the ACM* 30, no. 1 (January 1987): 32–47.

Strauss, Neil. "Rolling Stones Live on Internet: Both a Big Deal and a Little Deal." *New York Times*, November 22, 1994.

Streitz, Norbert. "An Interview with Ubicomp Pioneer Norbert Streitz." *IEEE Pervasive Computing* 11, no. 1 (January–March 2012): 62–66.

Suchman, Lucy. "Agencies at the Interface: Expanding Frames and Accountable Cuts." Lecture. *CHI Conference on Human Factors in Computing Systems*, April 13, 2010, Atlanta, GA.

———. "Anthropological Relocations and the Limits of Design." *Annual Review of Anthropology* 40, no. 1 (2011): 1–18.

———. "Conversation with Lucy Suchman." Interview by Claus Scharmer.

Presencing Institute, August 13, 1999. https://www.presencing.org/assets/images/aboutus/theory-u/leadership-interview/doc_suchman-1999.pdf.

———. *Plans and Situated Actions: The Problem of Human-Machine Communication*. New York: Cambridge University Press, 1987.

———. "Working Relations of Technology Production and Use." *Computer Supported Cooperative Work* 2 (March 1993): 21–29.

Suchman, Lucy, and Brigitte Jordan. "Interactional Troubles in Face-to-Face Survey Interviews." *Journal of the American Statistical Association* 85, no. 409 (1990): 232–41.

Sullivan, Bob. "'Calm Computing' Creator Dies at 46." ZDNet.com, May 4, 1999. https://www.zdnet.com/article/calm-computing-creator-dies-at-46/.

Sundblad, Yngve. "UTOPIA: Participatory Design from Scandinavia to the World." Lecture. *Third History of Nordic Computing Conference*. October 18, 2010, Stockholm, Sweden.

Takayama, Leila. "The Motivations of Ubiquitous Computing: Revisiting the Ideas behind and beyond the Prototypes." *Personal and Ubiquitous Computing* 21(2017): 557–69.

Tesler, Lawrence G. "Networked Computing in the 1990s." *Scientific American* 265, no. 3 (September 1991): 86–93.

Tierney, T. F. "Toronto's Smart City: Everyday Life or Google Life?" *Architecture_MPS* 15, no. 1 (2019): 1–21.

Topol, Susan. "A History of MTS—30 Years of Computing Service." *Information Technology Digest*, published by the University of Michigan Information Technology Division (May 1996): 9–17.

Turner, Fred. *From Counterculture to Cyberculture: Stewart Brand, the Whole Earth Network, and the Rise of Digital Utopianism*. Chicago: University of Chicago Press, 2006.

Turow, Joseph. *The Voice Catchers: How Marketers Listen In to Exploit Your Feelings, Your Privacy, and Your Wallet*. New Haven, CT: Yale University Press, 2021.

Usher, Rod. "The Third Wave." *Time*, Winter 1997–1998, 112–20.

Van, Jon. “Invisibility May Be Next for Technology.” *Chicago Tribune*, June 14, 1992.

Victor, Bret. “A Few Words on Doug Engelbart.” Worrydream.com, July 3, 2013. http://worrydream.com/Engelbart/.

Wagner, Mitch. “HAL Is Born.” *Computerworld* (January 6, 1997): 70–72.

Waldrop, M. Mitchell. *The Dream Machine: J. C. R. Licklider and the Revolution That Made Computing Personal*. Sloan Technology Series. New York: Viking, 2001.

Waldrop, M. Mitchell. “PARC Builds a World Saturated with Competition.” *Science* 261, no. 5128 (September 17, 1993): 1523–24.

Want, Roy, Bill N. Schilit, Norman I. Adams, Rich Gold, Karin Petersen, David Goldberg, John R. Ellis, and Mark Weiser. “An Overview of the PARCTAB Ubiquitous Computing Experiment.” *IEEE Personal Communications* 2, no. 6 (December 1995): 28–43.

Wasserman, Elizabeth. “Here, There and Everywhere.” *San Jose Mercury News*, January 6, 1998.

Weiser, Mark. “The Computer for the 21st Century.” *Scientific American* 265, no. 3 (September 1991): 94–104.

———. “Datorn Försvinner: Interview with Mark Weiser.” By Hans Sandberg. *DataTeknik*, March 1992.Weiser, Mark. “The Future of Ubiquitous Computing on Campus.” *Communications of the ACM* 41, no. 1 (January 1998): 41–42.

———. “Mark and Nicole’s Greek Vacation.” Mark Weiser’s personal website. August 16, 1998. https://web.archive.org/web/20161201013729/http:/www.ubiq.com/weiser/greece/.

———. “Open House.” *ITP Review* 2.0, March 1996. https://calmtech.com/papers/open-house.html.

———. “The Origins of Ubiquitous Computing Research at PARC in the Late 1980s.” *IBM Systems Journal* 38, no. 4 (1999): 693–96.

———. “The Spirit of the Engineering Quest.” *Technology in Society* 21, no. 4 (1999): 355–61.

———. “The Technologist’s Responsibilities and Social Change.” *Computer-Mediated Communication Magazine* 2, no. 4 (April 1, 1995): 17.

———. "Ubiquitous Computing." *IEEE Computer* 26, no. 10 (October 1993): 71–72.

———. "The World Is Not a Desktop." *ACM Interactions* (January 1994): 7–8.

Weiser, Mark, and John Seely Brown. "Center and Periphery: Balancing the Bias of Digital Technology." In *Blueprint to the Digital Economy*, edited by Don Tapscott, Alex Lowy, and David Ticoll, 317–36. New York: McGraw-Hill, 1998.

———. "The Coming Age of Calm Technology." In *Beyond Calculation: The Next Fifty Years of Computing*, edited by Peter J. Denning and Robert M. Metcalfe. New York: Copernicus, 1997.

———. "Designing Calm Technology." Report. Xerox PARC. December 21, 1995.

"Welcome to the University of Michigan Computing Center." Unpublished manual, January 1978. http://archive.michigan-terminal-system.org/discussions/u-m-computing-center.

Wells, H. G. *World Brain*. Cambridge, MA: MIT Press, 2021.

"Where the Future Is Being Designed." *Cronaca*, July 1992.

Wolkomir, Richard. "We're Going to Have Computers Coming Out of the Woodwork." *Smithsonian* 25, no. 6 (September 1994): 82–90.

Wright, Sarah H. "'Wearables' Combine Fashion, High-Tech." *MIT News*, October 22, 1997. https://news.mit.edu/1997/wearables-1022.

Xerox. "Xerox Names Computing Pioneer as Chief Technologist for Palo Alto Research Center." News release. August 14, 1996.

"Xerox PARC Has New Tech Leader." *San Francisco Examiner*, August 15, 1996.

Zakon, Robert H. "Hobbes' Internet Timeline 25." Zakon.org, January 1, 2018. https://www.zakon.org/robert/internet/timeline/.

Zuboff, Shoshana. *The Age of Surveillance Capitalism: The Fight for a Human Future at the New Frontier of Power*. New York: PublicAffairs, 2018.

———. "Toronto Is Surveillance Capitalism's New Frontier." *Toronto Life*, September 4, 2019. https://torontolife.com/city/toronto-is-surveillance-capitalisms-new-frontier/.

INDEX

Note: Page numbers in italics refer to figures.